T0313022

Opsin-free Optogenetics

Optogenetics represents a breakthrough technology capable of dynamically modulating molecular and cellular activity in live cells with high precision. This transformative technology made it possible to reversibly interrogate protein actions, cellular events, and animal behaviors with a simple flash of light. Manipulating the temporal and spatial profile of light enabled precise control of membrane potentials. This feature has inspired non-opsin-based optogenetics, which not only inherits the precise spatiotemporal resolution but also expands the targets into a diverse pool of biomolecules such as membrane receptors, ion channels, kinases, GTPases, and transcription factors. This book is unique because it emphasizes the design and applications of opsin-free optogenetic tools.

Key Features

- Describes new developments in non-opsin-based optogenetic tools
- Introduces cutting-edge methods for precise modulation of cell signaling
- Highlights applications of optogenetic regulation of cellular functions
- Covers contributions from an international team of leading experts

Methods in Signal Transduction

Series Editors: Joseph Eichberg, Jr. and Michael X. Zhu

The overall theme of this series continues to be the presentation of the wealth of up-to-date research methods applied to the many facets of signal transduction. Each volume is assembled by one or more editors who are pre-eminent in their specialty. In turn, the guiding principle for editors is to recruit chapter authors who will describe procedures and protocols with which they are intimately familiar in a reader-friendly format. The intent is to assure that each volume will be of maximum practical value to a broad audience, including students and researchers just entering an area, as well as seasoned investigators.

Calcium Entry Channels in Non-Excitable Cells
Juliusz Ashot Kozak and James W. Putney, Jr.

Autophagy and Signaling
Esther Wong

Signal Transduction and Smooth Muscle
Mohamed Trebak and Scott Earley

Polycystic Kidney Disease
Jinghua Hu and Yong Yu

New Techniques for Studying Biomembranes
Qiu-Xing Jiang

Ion and Molecule Transport in Lysosomes
Bruno Gasnier and Michael X. Zhu

Neuron Signaling in Metabolic Regulation
Qingchu Tong

Non-Classical Ion Channels in the Nervous System
Tian-Le Xu, Long-Ju Wu

Immune cells, Inflammation and Cardiovascular Diseases
Shyam S. Bansal

Opsin-free Optogenetics: Technology and Applications
Yubin Zhou and Kai Zhang

For more information about this series, please visit: www.crcpress.com/Methods-in-Signal-Transduction-Series/book-series/CRCMETSIGTRA?page=&order=pubdate&size=12&view=list&status=published,forthcoming

Opsin-free Optogenetics

Technology and Applications

Edited by Yubin Zhou and Kai Zhang

CRC Press
Taylor & Francis Group
Boca Raton London New York

CRC Press is an imprint of the
Taylor & Francis Group, an **informa** business

First edition published 2023
by CRC Press
6000 Broken Sound Parkway NW, Suite 300, Boca Raton, FL 33487–2742

and by CRC Press
4 Park Square, Milton Park, Abingdon, Oxon, OX14 4RN

CRC Press is an imprint of Taylor & Francis Group, LLC

Library of Congress Cataloging-in-Publication Data
A catalog record for this book has been requested

ISBN: 978-1-032-24922-3 (hbk)
ISBN: 978-1-032-24923-0 (pbk)
ISBN: 978-1-003-28075-0 (ebk)

DOI: 10.1201/b22823

Typeset in Times
by Apex CoVantage, LLC

Content

Preface....................vii
Acknowledgments....................ix
Editor Bios....................xi
List of Contributors....................xiii

Chapter 1 Optogenetic Dissection of a Two-Component Calcium Influx Pathway....................1

Xiaoxuan Liu, Yingshan Wang, Yubin Zhou, and Guolin Ma

Chapter 2 High-Throughput Engineering of a Light-Activatable Ca^{2+} Channel....................25

Lian He, Liuqing Wang, and Youjun Wang

Chapter 3 Optogenetic Activation of TrkB Signaling....................37

Peiyuan Huang, Zhihao Zhao, Lei Lei, and Liting Duan

Chapter 4 Spatiotemporal Modulation of Neural Repair....................55

Huaxun Fan, Qin Wang, Kai Zhang, and Yuanquan Song

Chapter 5 Optogenetic Control of Neural Stem Cell Differentiation....................71

Yixun Su, Taida Huang, Kai Zhang, and Chenju Yi

Chapter 6 An Optogenetic Toolbox for Remote Control of Programmed Cell Death....................85

Ningxia Zhang and Ji Jing

Chapter 7 Control of Protein Levels in *Saccharomyces cerevisiae* by Optogenetic Modules that Act on Protein Synthesis and Stability....................99

Sophia Hasenjäger, Jonathan Trauth, and Christof Taxis

Chapter 8 Optogenetics as a Tool to Study Neurodegeneration and Signal Transduction....................111

Prabhat Tiwari and Nicholas S. Tolwinski

Chapter 9 Opsin-free Optogenetics: Brain and Beyond.................................... 123

Jongryul Hong, Yeonji Jeong, and Won Do Heo

Chapter 10 Constructing a Far-Red Light-Induced Split-Cre Recombinase System for Controllable Genome Engineering 135

Meiyan Wang, Jiali Wu, and Haifeng Ye

Chapter 11 Tools and Technologies for Wireless and Non-Invasive Optogenetics.. 151

Guangfu Wu, Vagif Abdulla, Yiyuan Yang, Michael J. Schneider, and Yi Zhang

Index 179

Preface

Following the first ectopic introduction of *Chlamydomonas reinhardtii* channel-rhodopsin 2 (ChR2) into mammals to enable optical control of neurons and neural circuits in 2005, the optogenetics field has witnessed an exponential growth with annual publication numbers reaching over 1,000 in the past five years. Microbial opsin-based optogenetics has been nothing short of revolutionary for the neuroscience field. Optogenetics was selected as the "Method of the Year 2010" by Nature Methods. Pioneers in this field were honored with the Albert Lasker Basic Medical Research Award 2021. Owing to the radical difference in the localization and excitability between excitable and non-excitable cells or tissues, a parallel evolution of optogenetic tools tailored for these biological systems requires quite different design principles. Recently, a series of non-opsin-based photoswitches or photosensitive modules, derived from plants and microbes, have gained wide applications in the remote control of protein activity, cellular physiology, and animal behaviors. These non-opsin-based molecular tools usher in the next stage of optogenetics.

This volume of the CRC series on Methods in Signal Transduction contains 11 chapters that cover a broad range of optogenetic engineering targets, including ion channels (e.g., calcium release-activated calcium channel), membrane receptors and their downstream effectors (e.g., receptor tyrosine kinases and downstream ERK and PI3K signaling components), programmed cell death pathways (apoptosis, necroptosis, and pyroptosis), the proteasomal degradation pathway, neurotoxic proteins, DNA recombinases, and the CRISPR/Cas-based genome engineering machinery. This book also highlights innovative light-delivery approaches to enable wireless and remote control of living organisms by using two-photon excitation, bioluminescence, upconversion nanoparticles, and wearable miniature microLED devices.

We would like to thank Dr. Michael Zhu for the invitation to curate this book and all contributing authors for their great efforts. We also greatly appreciate the guidance and hard work from the editorial team at CRC Press, including Chuck Crumly and Kara Roberts, and from the production team at Apex CoVantage, including Aruna Rajendran, to accelerate publication of this book. This work is dedicated to the 2008 Nobel laureate Dr. Roger Tsien, who has been an inspiring mind and a giant in the field of protein engineering and whose work has led to paradigm-shifting advances in the design of biosensors and actuators over the last three decades.

Acknowledgments

Thanks to Chuck Crumly, Kara Roberts, and Michael X. Zhu for all the support along the way. It surely is a rewarding experience.

Editor Bios

Yubin Zhou. Dr. Yubin Zhou is currently a professor, Presidential Impact Fellow, and American Cancer Society Research Scholar in the Institute of Biosciences and Technology at Texas A&M University. He is also a faculty in the Department of Translational Medical Sciences, School of Medicine at Texas A&M. Dr. Zhou received his medical training in internal medicine from Zhejiang University School of Medicine. He thereafter earned his MSci degree in chemistry and bioinformatics (2007), as well as his PhD degree in chemistry and virology (2008) from Georgia State University. After receiving his postdoctoral training as a Leukemia & Lymphoma Society Fellow at Harvard Medical School (2008–2010) and later working as an instructor at La Jolla Institute for Immunology (2010–2012), Dr. Zhou launched a vibrant research program centered on synthetic immunology and bioengineering at Texas A&M University.

A tight integration among mechanistic studies, biomedical engineering, and translational sciences is a hallmark of Dr. Zhou's research. His group is surfing at the frontier of three research areas: (1) pioneering chemical and synthetic biology approaches for precise and programmable control of protein activity and cell physiology; (2) illuminating novel regulatory mechanisms of signal transduction that remain unresolved in calcium signaling and T cell activation or exhaustion; and (3) developing conditionally active biologics and cell-based therapies for cancer treatment and neuromodulation.

Dr. Zhou has published over 140 peer-reviewed research papers and organized several prominent national and international conferences in calcium signaling and optogenetics. He currently serves as the associate editor for *Current Molecular Medicine*, and sits on the editorial boards of *Molecular and Cellular Biology*, *Chronic Disease and Translational Medicine*, and *Frontiers in Molecular Biosciences*. Molecular tools created by the Zhou lab have been widely distributed to 200+ research laboratories across the globe.

Kai Zhang. Dr. Kai Zhang is Associate Professor and Scialog Fellow in the Department of Biochemistry at the University of Illinois at Urbana-Champaign (UIUC). He is an affiliate faculty to the Beckman Institute, Neuroscience Program, Center for Biophysics and Quantitative Biology, Cancer Center at Illinois at UIUC. Dr. Zhang received his Bachelor of Science from the University of Science and Technology of China (USTC) in 2002 and a PhD in chemistry from the University of California, Berkeley in 2008. During training at Stanford University as an American Cancer Society postdoctoral fellow, he made a transition from the field of physical chemistry to neurobiology. In August 2014, Dr. Zhang joined the Biochemistry Department of UIUC as a tenure-track assistant professor and was promoted to Associate Professor in 2021.

At Illinois, Dr. Zhang has established a multi-disciplinary research program by integrating his expertise in biochemistry, biophysics, bioengineering, and neuroscience. The Zhang lab uses a synthetic biology approach to understand how physiological signaling processes are compromised in diseases and develops enabling biotechnologies to restore impaired cell functions. The key focus of his research program includes (1) promoting neural regeneration after injury, (2) dynamic controlling of cell fate determination during embryonic development, (3) establishing generalizable optogenetic strategies for post-translational control of protein activity, and (4) developing biosensors with new nanomaterials.

To date, Dr. Zhang has published 60 peer-reviewed research articles. Every year, he enjoys teaching Physical Biochemistry to junior and senior undergraduate students and first-year graduate students. He received the Innovative Teaching and Learning Grant to promote interactive learning in the classroom. He serves as the Associate Editor of the *Journal of Molecular Biology* and *Frontiers in Molecular Neuroscience*. Prior to this book, Dr. Zhang edited a special issue on chemo- and optogenetics for the *Journal of Molecular Biology*.

Contributors

Vagif Abdulla
Department of Biomedical Engineering and the Institute of Materials Science
University of Connecticut
Storrs, Connecticut

Won Do Heo
Department of Biological Sciences and Institute for the BioCentury
Korea Advanced Institute of Science and Technology (KAIST)
Daejeon, Republic of Korea

Liting Duan
Department of Biomedical Engineering
The Chinese University of Hong Kong
Sha Tin, Hong Kong SAR, China

Huaxun Fan
Department of Biochemistry
University of Illinois at Urbana-Champaign
Urbana, Illinois

Sophia Hasenjäger
Department of Biology/Genetics
Philipps-University Marburg
Marburg, Germany

Lian He
Department of Pharmacology
School of Medicine
Southern University of Science and Technology
Shenzhen, Guangdong, China

Jongryul Hong
Department of Biological Sciences
Korea Advanced Institute of Science and Technology (KAIST)
Daejeon, Republic of Korea

Peiyuan Huang
Department of Biomedical Engineering
The Chinese University of Hong Kong
Sha Tin, Hong Kong SAR, China

Taida Huang
The Seventh Affiliated Hospital of Sun Yat-sen University
Shenzhen, Guangdong, China

Yeonji Jeong
Department of Biological Sciences
Korea Advanced Institute of Science and Technology (KAIST)
Daejeon, Republic of Korea

Ji Jing
Key Laboratory of Prevention, Diagnosis and Therapy of Upper Gastrointestinal Cancer of Zhejiang Province
The Cancer Hospital of the University of Chinese Academy of Sciences and Institute of Basic Medicine and Cancer Chinese Academy of Sciences
Hangzhou, China

Lei Lei
Department of Biomedical Engineering
City University of Hong Kong
Kowloon, Hong Kong SAR, China

Xiaoxuan Liu
Center for Translational Cancer Research
Institute of Biosciences and Technology
Texas A&M University
Houston, Texas

Guolin Ma
Center for Translational Cancer Research
Institute of Biosciences and Technology
Texas A&M University
Houston, Texas

Nicholas S. Tolwinski
Division of Science
Yale-NUS College Singapore

Michael J. Schneider
Department of Biomedical Engineering and the Institute of Materials Science
University of Connecticut
Storrs, Connecticut

Yuanquan Song
Raymond G. Perelman Center for Cellular and Molecular Therapeutics
The Children's Hospital of Philadelphia
Department of Pathology and Laboratory Medicine
University of Pennsylvania
Philadelphia, Pennsylvania

Yixun Su
The Seventh Affiliated Hospital of Sun Yat-sen University
Shenzhen, Guangdong, China

Christof Taxis
School of Life Sciences
Siegen University
Siegen, Germany

Prabhat Tiwari
NYU Langone Medical Center New York City
New York, New York

Jonathan Trauth
Department of Biology/Genetics
Philipps-University Marburg
Marburg, Germany

Liuqing Wang
Beijing Key Laboratory of Gene Resource and Molecular Development
College of Life Sciences
Beijing Normal University
Beijing, China

Meiyan Wang
Biomedical Synthetic Biology Research Center
Shanghai Key Laboratory of Regulatory Biology
Institute of Biomedical Sciences and School of Life Sciences
East China Normal University
Shanghai, China

Qin Wang
Raymond G. Perelman Center for Cellular and Molecular Therapeutics
The Children's Hospital of Philadelphia
Department of Pathology and Laboratory Medicine
University of Pennsylvania
Philadelphia, Pennsylvania

Yingshan Wang
Center for Translational Cancer Research Institute of Biosciences and Technology
Texas A&M University
Houston, Texas

Youjun Wang
Beijing Key Laboratory of Gene Resource and Molecular Development
College of Life Sciences
Beijing Normal University
Beijing, China

Guangfu Wu
Department of Biomedical Engineering and the Institute of Materials Science
University of Connecticut
Storrs, Connecticut

Jiali Wu
Biomedical Synthetic Biology Research Center
Shanghai Key Laboratory of Regulatory Biology
Institute of Biomedical Sciences and School of Life Sciences
East China Normal University
Shanghai, China

Yiyuan Yang
Department of Mechanical Engineering
Massachusetts Institute of Technology
Cambridge, Massachusetts

Haifeng Ye
Biomedical Synthetic Biology Research Center
Shanghai Key Laboratory of Regulatory Biology
Institute of Biomedical Sciences and School of Life Sciences
East China Normal University
Shanghai, China

Chenju Yi
The Seventh Affiliated Hospital of Sun Yat-sen University
Shenzhen, Guangdong, China

Kai Zhang
Department of Biochemistry
University of Illinois at Urbana-Champaign
Urbana, Illinois

Ningxia Zhang
The Cancer Hospital of the University of Chinese Academy of Sciences
Institute of Basic Medicine and Cancer
Chinese Academy of Sciences and Department of Medical Oncology
Sir Run Run Shaw Hospital, College of Medicine
Zhejiang University
Hangzhou, Zhejiang, China

Yi Zhang
Department of Biomedical Engineering and the Institute of Materials Science
University of Connecticut
Storrs, Connecticut

Yubin Zhou
Center for Translational Cancer Research
Institute of Biosciences and Technology
Texas A&M University
Houston, Texas

Zhihao Zhao
Department of Biomedical Engineering
The Chinese University of Hong Kong
Sha Tin, Hong Kong SAR, China

1 Optogenetic Dissection of a Two-Component Calcium Influx Pathway

Xiaoxuan Liu, Yingshan Wang, Yubin Zhou, and Guolin Ma

CONTENTS

1.1 Introduction 1
1.2 Overview of the Optogenetic Toolbox 2
1.3 Optogenetic Approaches to Dissecting Signaling Pathways 4
1.3.1 Optogenetic Mimicry of Molecular Steps Involved in SOCE 5
1.3.2 CRAC Channel-Based Optogenetics for Versatile Control of Ca^{2+} Signaling 5
1.3.3 Optogenetic Clustering to Map Out Domains Required for PPIs *in cellulo* 7
1.3.4 Optogenetic Dissection of Protein Self-Oligomerization and PPIs 10
1.3.5 Dissecting STIM1-Organelle Interactions with Optogenetic Approaches 13
1.3.6 Optogenetic Dissection of the Functional Differences in STIM Variants 14
1.4 Conclusions 17
Acknowledgment 17
References 17

1.1 INTRODUCTION

Optogenetics is a powerful technique that combines genetics with optics to remotely regulate protein actions and control biological functions in living cells, tissues, and organisms[1–4]. To date, optogenetic applications have achieved remarkable successes in multiple fields, including neuroscience, immunology, cell biology, and developmental biology, by conferring tight control over different cell signaling pathways[2,5]. Multiple opsin-free photosensitive proteins originating from plants, bacteria, and fungi have been extensively engineered to manipulate protein conformational changes, dimerization or oligomerization, subcellular localization, and protein-protein interactions (PPIs). Optogenetics is well known for its non-invasiveness,

DOI: 10.1201/b22823-1

rapid responsiveness, superior reversibility, and high spatiotemporal precision[2,6]. Optogenetic approaches have greatly facilitated the dissection of the structure-function relationship of cell signaling proteins at the molecular level[7–13], as best exemplified by the store-operated calcium entry (SOCE) pathway.

SOCE serves as a major route for Ca^{2+} entry in many cell types, with the Ca^{2+} release-activated Ca^{2+} (CRAC) channel composed of ORAI and stromal interaction molecule (STIM) as the most well-known prototype of SOCE[14–19]. Tremendous progress has been made regarding the molecular steps involved in SOCE activation over the past decade[14–19]. SOCE is initiated upon Ca^{2+} release from intracellular Ca^{2+} stores, such as endoplasmic/sarcoplasmic reticulum ER/SR[19,20], followed by STIM1 oligomerization and migration toward the plasma membrane (PM) to engage and gate the ORAI Ca^{2+} channels. ORAI Ca^{2+} channels mediate the influx of extracellular Ca^{2+} into the cytosol to further activate Ca^{2+}-responsive phosphatase calcineurin and ultimately drive the nuclear entry of nuclear factor of activated-T cells (NFAT)[15–19]. The CRAC channel-based optogenetic toolbox (designated Opto-CRAC), recently developed by us and others, provides exciting new opportunities to remotely and reversibly control Ca^{2+} signaling and modulate Ca^{2+}-dependent physiological processes with high precision[21–23].

This chapter focuses on explaining the principles of optogenetic engineering applied to control different steps and cellular events during SOCE, highlighting how STIM1-inspired optogenetic tools can be exploited to dissect the mechanisms involved in SOCE activation[21, 24–30]. This optogenetic engineering approach can be broadly extended to facilitate the mechanistic dissection of other important cell signaling pathways as well.

1.2 OVERVIEW OF THE OPTOGENETIC TOOLBOX

Building upon photoreceptors derived from plants, bacteria, or fungi, a series of non-opsin-based photoswitchable modules have been developed and engineered to control protein activity by light at varying wavelengths[2]. Based on their working mechanisms, these photosensitive modules can be divided into three categories (Figure 1.1): light-sensitive allosteric switches, photo-inducible dimerization or oligomerization systems, and photo-inducible dissociation systems[2].

Light-sensitive allosteric switch modules. The light-oxygen-voltage (LOV) domains derived from plants are among the most widely used single-component photosensory modules for allosteric regulation of a protein of interest (POI)[31–33]. In a typical design, an effector domain from POIs is fused to the C-terminus of the LOV2 domain derived from *Avena sativa* (AsLOV2), anticipating to blocking the active site of POI via steric hindrance in the dark[2, 34–39]. Upon photostimulation, the C-terminal Jα helix of AsLOV2 undergoes conformational changes to release the fused effector domain, thereby restoring its activity. To further expand the caging interfaces of LOV2, circular permutation has been applied to yield cpLOV2 to afford additional caging capability via fusion of the effector domain to the N-terminus of cpLOV2 or between the re-ordered Jα helix and the PAS core

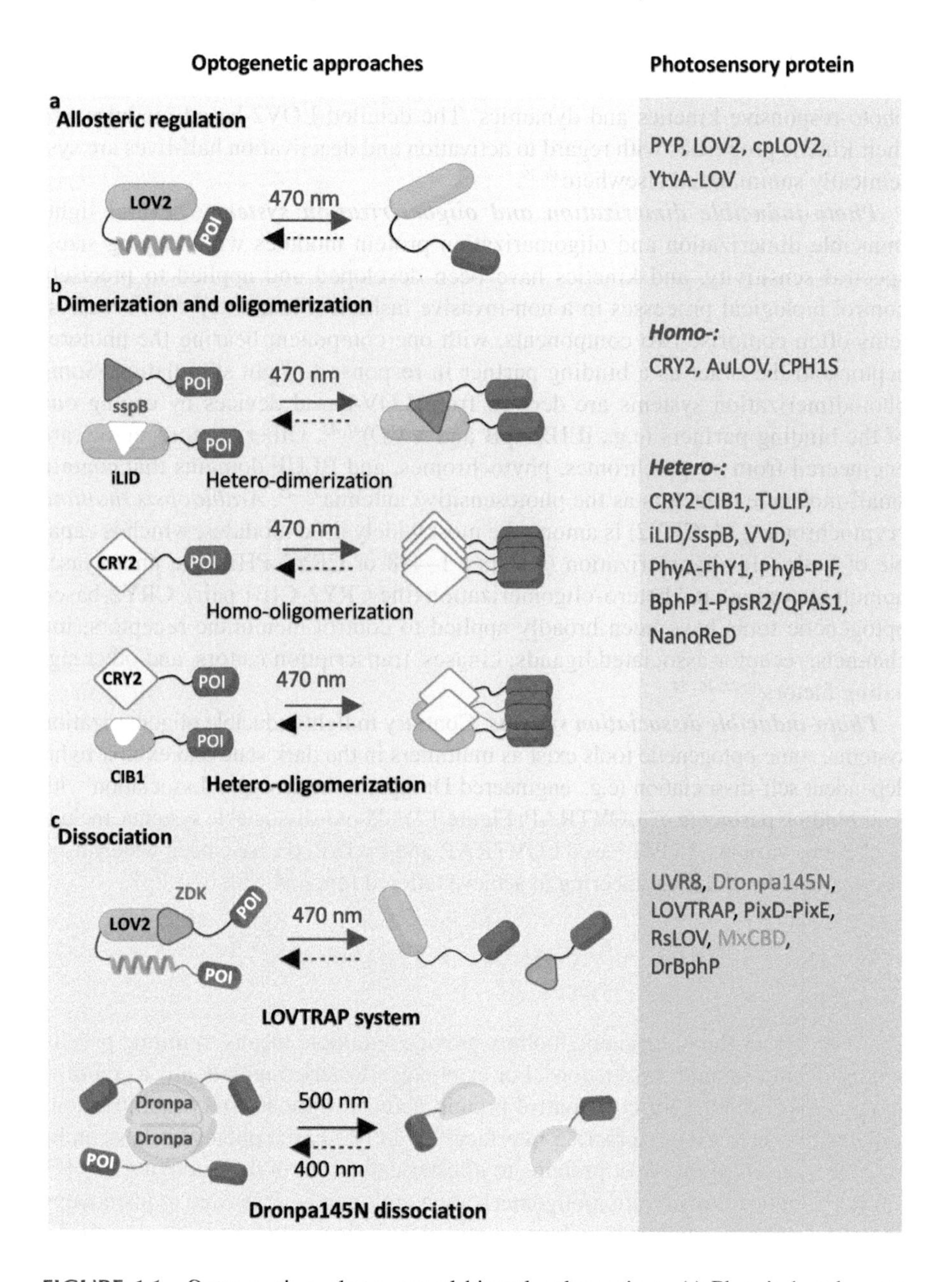

FIGURE 1.1 Optogenetic tools to control biomolecular actions. (a) Photoinduced uncaging of the fused protein of interest (POI) by Jα helix unfolding in AsLOV2 or cpLOV2. (b) Light-induced dimerization and oligomerization platforms including homo-dimerization/oligomerization and hetero-dimerization/oligomerization. (c) Photoinduced dissociation system.

domain of LOV2[2, 29]. Moreover, multiple fast and slow cycling mutations have been identified and introduced into LOV2-based optogenetic devices to diversify their photo-responsive kinetics and dynamics. The detailed LOV2-based modules and their kinetic properties with regard to activation and deactivation half-lives are systemically summarized elsewhere[38, 39].

Photo-inducible dimerization and oligomerization systems. Several light-inducible dimerization and oligomerization protein modules with varying sizes, spectral sensitivity, and kinetics have been developed and applied to precisely control biological processes in a non-invasive fashion (Figure 1.1)[2, 40]. These systems often comprise two components, with one component bearing the photoreceptor and the other as a binding partner in response to light stimulation. Some photodimerization systems are derived from LOV-based devices by caging one of the binding partners (e.g., iLID/sspB and VVD)[41, 42]. Other photoreceptors are engineered from cryptochromes, phytochromes, and BLUF domains that contain small-molecule cofactors as the photosensitive antenna[43–45]. *Arabidopsis thaliana* cryptochrome 2 (AtCRY2) is among the most widely used modules, which is capable of both self-oligomerization (residues 1–498 or CRY2-PHR, the photolyase-homology region) and hetero-oligomerization (the CRY2-CIB1 pair). CRY2-based optogenetic tools have been broadly applied to control membrane receptors, ion channels, receptor-associated ligands, kinases, transcription factors, and other signaling factors[10, 11, 46–54].

Photo-inducible dissociation systems. Contrary to light-inducible oligomerization systems, some optogenetic tools exist as multimers in the dark state and exhibit light-dependent self-dissociation (e.g., engineered Dronpa) or heterotypic dissociation with their binding partner (e.g., LOVTRAP; Figure 1.1)[2]. Photo-dissociable systems, including Dronpa variants, LOV2-based LOVTRAP, and PixD/PixE, have been widely used in protein and cellular engineering to achieve tailored function with light[55–59].

1.3 OPTOGENETIC APPROACHES TO DISSECTING SIGNALING PATHWAYS

The diversity of the optogenetic toolbox provides multiple means to mimic protein actions during signal transduction[2]. For example, allosteric regulation is a common strategy evolved by nature to control protein activity without direct modifications to the active site or the interacting interface[60, 61]. LOV2-based photoswitches can be modularly inserted into host proteins to allosterically control their activity[31, 38, 39, 62]. Similarly, optical dimerization/oligomerization tools can be employed to phenocopy oligomerization-dependent signaling events, such as ligand-induced receptor dimerization[2], in a light-dependent manner. As such, optogenetic engineering offers a non-conventional and non-invasive approach and complements existing pharmacological and genetic methods to facilitate the mechanistic dissection of complicated signaling networks[62, 63]. Herein, we dissect the signaling pathway of STIM1-ORAI-mediated SOCE by incorporating optogenetic tools into SOCE signaling components. We harness the power of light to control different signaling activities, including protein conformational switch, protein self-clustering, and protein-protein or protein-organelles interactions.

1.3.1 Optogenetic Mimicry of Molecular Steps Involved in SOCE

As an ER-resident Ca^{2+} sensor protein, STIM1 contains an ER-luminal Ca^{2+}-sensing domain (EF-SAM), a single transmembrane domain (TM), and a cytoplasmic domain (STIM1ct) that directly gates ORAI Ca^{2+} channels in the PM[14–19] (Figure 1.2a). The luminal EF-hand motif senses Ca^{2+} fluctuation in ER and cooperates with the sterile alpha motif domain (SAM) to transduce the Ca^{2+} signal from ER lumen to the cytoplasm[64,65]. Upon ER Ca^{2+} store depletion triggered by external ligand binding or antigen engagement of membrane receptors, Ca^{2+} dissociates from the EF-hand to cause EF-SMA oligomerization within the ER lumen[64–66]. The ER luminal signal is transduced through the TM domain toward the cytoplasmic domain that overcomes the intramolecular autoinhibition within STIM1ct mediated by coiled-coil region 1 (CC1) and the STIM Orai-activating domain (SOAR or CAD)[67,68]. The rearrangement of CC1 exposes SOAR and switches STIM1ct from a "folded-back" inactive state to a more extended active conformation[69–71]. Subsequently, activated STIM1 oligomers migrate toward regions of ER—PM appositions (termed "puncta") to gate ORAI1 channels through direct physical contacts[72,73]. This step-wise activation process is made possible via coordinated actions of various STIM1 cytoplasmic regions, including SOAR/CAD ($STIM1_{342–448}$), the S/TxIP EB1-binding motif (TRIP: $STIM1_{642–645}$), and the polybasic domain (PB: $STIM1_{671–685}$), which are involved in direct activation of ORAI channels, interaction with microtubule (MT) associated +TIP tracking protein EB1, and binding to PM-resident phosphoinositides (PIPs), respectively (Figure 1.2b)[37,74–78].

Mimicking the oligomerization and conformational switch steps during STIM1 activation, various engineering efforts have been made to recapitulate STIM1-mediated signaling events (Figure 1.2c): (1) CC1 is replaced by LOV2 to reversibly mimic CC1-SOAR mediated autoinhibition with light [24,25,29]; (2) the luminal EF-SAM domain is substituted with CRY2 to control STIM1 activation[7,26]; and (3) CRY2 is fused to the TRIP and PB motifs to manipulate microtubule tracking and protein targeting to the PM in a light-dependent manner[7,79].

Optogenetic approaches can also fully mimic the function of the full-length STIM1 activation, ER-PM contact site assembly, and Ca^{2+} microdomain formation[7]. By tethering CRY2 fused STIM1ct on the ER membrane, light illumination leads to the formation of puncta-like structures and Ca^{2+} influx at the ER-PM junction. Because of its superior reversibility, CRY2-STIM1ct was used to demonstrate that de-oligomerization of STIM1ct could bring STIM1 back to its resting configuration, even in the absence of other ancillary proteins.

1.3.2 CRAC Channel-Based Optogenetics for Versatile Control of Ca^{2+} Signaling

A variety of engineering strategies have been adopted to craft a suite of genetically encoded Ca^{2+} actuators (GECAs) based on CRAC channels[21–23]. CRY2 and LOV2-based optogenetic modules have been engineered into different STIM1 fragments and tethered to subcellular organelles to yield Opto-CRAC variants with

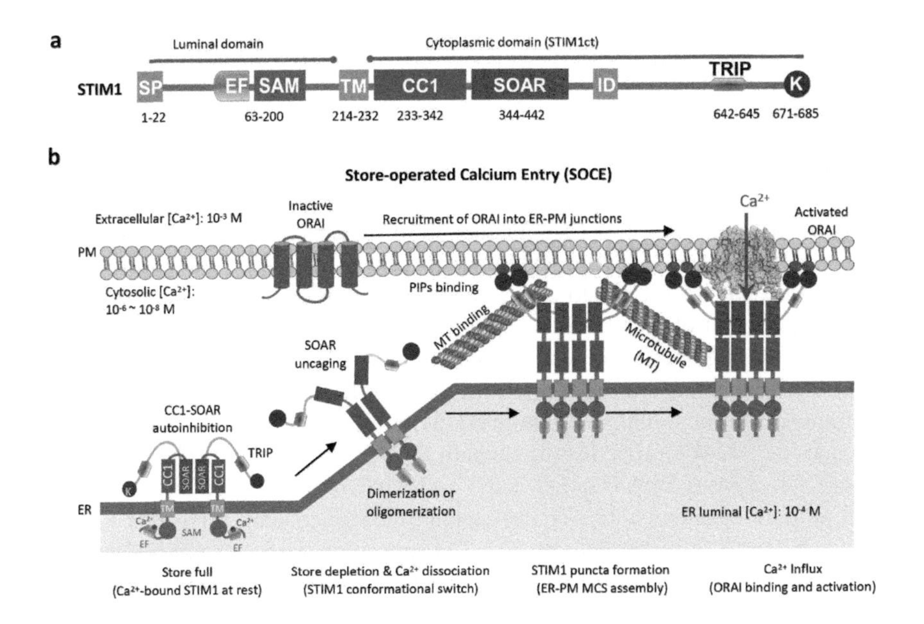

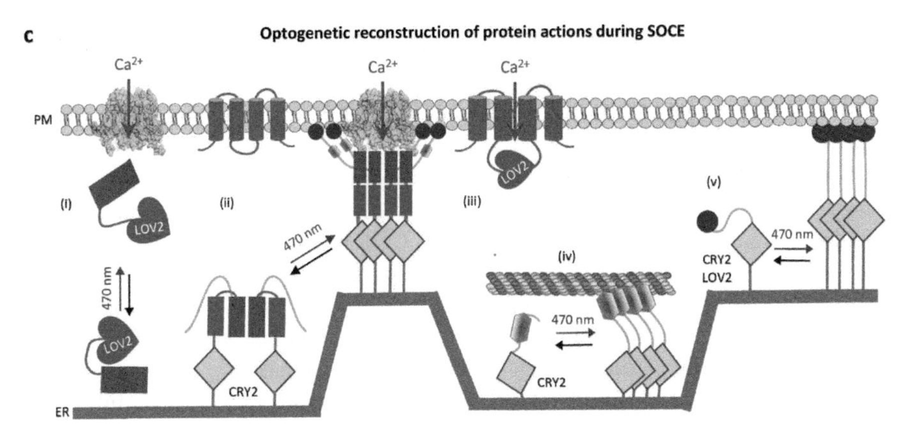

FIGURE 1.2 Optogenetic approaches to dissect the molecular choreography of SOCE mediated by ORAI-STIM coupling. (a) Domain architecture of human STIM1. SP, signal peptide; EF-SAM, EF-hand and sterile alpha-motif; TM, transmembrane domain; CC1, coiled-coil domain 1; SOAR, STIM-Orai activating region; P/S, proline/serine-rich region; TRIP, the S/TxIP microtubule-binding motif; PB, polybasic tail. (b) Functional STIM1-ORAI1 coupling to mediate SOCE. (c) Optogenetic mimicry of key molecular steps involved in SOCE, including (i) the use of LOV2 to cage SOAR (*e.g.*, Opto-CRAC1 variants, BACCS variants, LOVS1K); (ii) use of CRY2 to enable light-inducible oligomerization and activation of STIM1ct (Opto-CRAC2 or optoSTIM); (iii) direct insertion of LOV2 into ORAI1 channels to generate light-operated Ca^{2+} channel (LOCa); (iv) light-inducible STIM1-MT binding and ER-MT contact; and v) photo-inducible STIM1-phosphoinositides (on the PM) association, as well as ER-PM contact formation in a reversible manner.

different activation and deactivation profiles[7, 24–26, 28, 29, 80–82]. CRY2-based GECAs (OptoSTIM1 or Opto-CRAC2) have a relatively slower deactivation half-life compared to LOV2-based versions (Opto-CRAC1 or BACCS2) (Figure 1.3b)[7, 24–26, 28, 29, 81], making them an ideal choice to produce sustained and stronger Ca^{2+} signals with a single pulse of light. To mimic physiologically relevant Ca^{2+} oscillations in mammalian cells, LOV2-based GECAs seem to be a better choice given their relatively faster kinetics[24]. These photo-responsive GECAs make it possible to generate customized Ca^{2+} oscillation patterns to affect downstream NFAT-dependent gene expression in mammals by varying the amplitude and frequency of light inputs (Figure 1.3c)[24, 27].

In contrast to the light-ON devices described previously, light can also be utilized to shut off Ca^{2+} flux into the cytosol. This can be achieved by installing the Zdk2—cpLOV2 pair into a STIM1ct fragment (residues 233–473, CC1-SOAR)[29]. In the dark, the SOAR domain remains exposed to gate ORAI channels and elicits constitutive Ca^{2+} influx (Figure 1.3a). Following exposure to blue light, the Zdk2—cpLOV2 association is disrupted to restore the CC1—SOAR autoinhibition, thereby terminating Ca^{2+} entry.

Optogenetic engineering has also been used on the pore-forming subunit of the CRAC channel, ORAI1, to generate a light-operated Ca^{2+} channel (LOCa) (Figure 1.2c and 1.3a)[30]. LOCa was designed by inserting LOV2 into the intracellular loop of a constitutively active human ORAI1 variant bearing the mutations H171D and P245T[83]. Complementary to the Opto-CRAC series derived from STIM1, LOCa enables optical control over Ca^{2+} signals and Ca^{2+}-dependent hallmark physiological responses independent of endogenous ORAI and STIM expression. The activation half-life of LOCa is slightly slower than STIM1-based tools, but the deactivation half-life falls between those of CRY2-based and LOV2-based GECAs. LOCa may be used as an ideal tool for the control of physiological processes in cells or tissues with low endogenous ORAI expression and having slower requirements on kinetics, such as gene expression, immunomodulation, and cell metabolism.

1.3.3 Optogenetic Clustering to Map Out Domains Required for PPIs *in Cellulo*

At the resting condition, STIM1 is kept in an inactive state presumably by adopting a folded-back conformation via concerted actions of three structural determinants: (1) the Ca^{2+}-bound ER luminal EF-SAM domain; (2) an intramolecular clamp mediated by cytosolic CC1-SOAR interaction; and (3) an inhibitory domain (ID; $STIM1_{473-491}$) downstream of SOAR[15, 70, 84]. When split at the boundary between CC1 and SOAR at position 342, the two STIM1 fragments (1–342 and 343–685) still stay together based on colocalization in live-cell fluorescence microscopy images (Figure 1.4a)[7], providing compelling evidence to support interaction between CC1 and SOAR. To further validate the intramolecular trapping model of STIM1 autoinhibition mediated by CC1-SOAR, the CC1 domain was replaced by LOV2 to yield LOV2-SOAR variants, anticipating caging SOAR to prevent constitutive activation of ORAI channels. The inclusion of ID domain in the LOV2-SOAR chimera was found to achieve

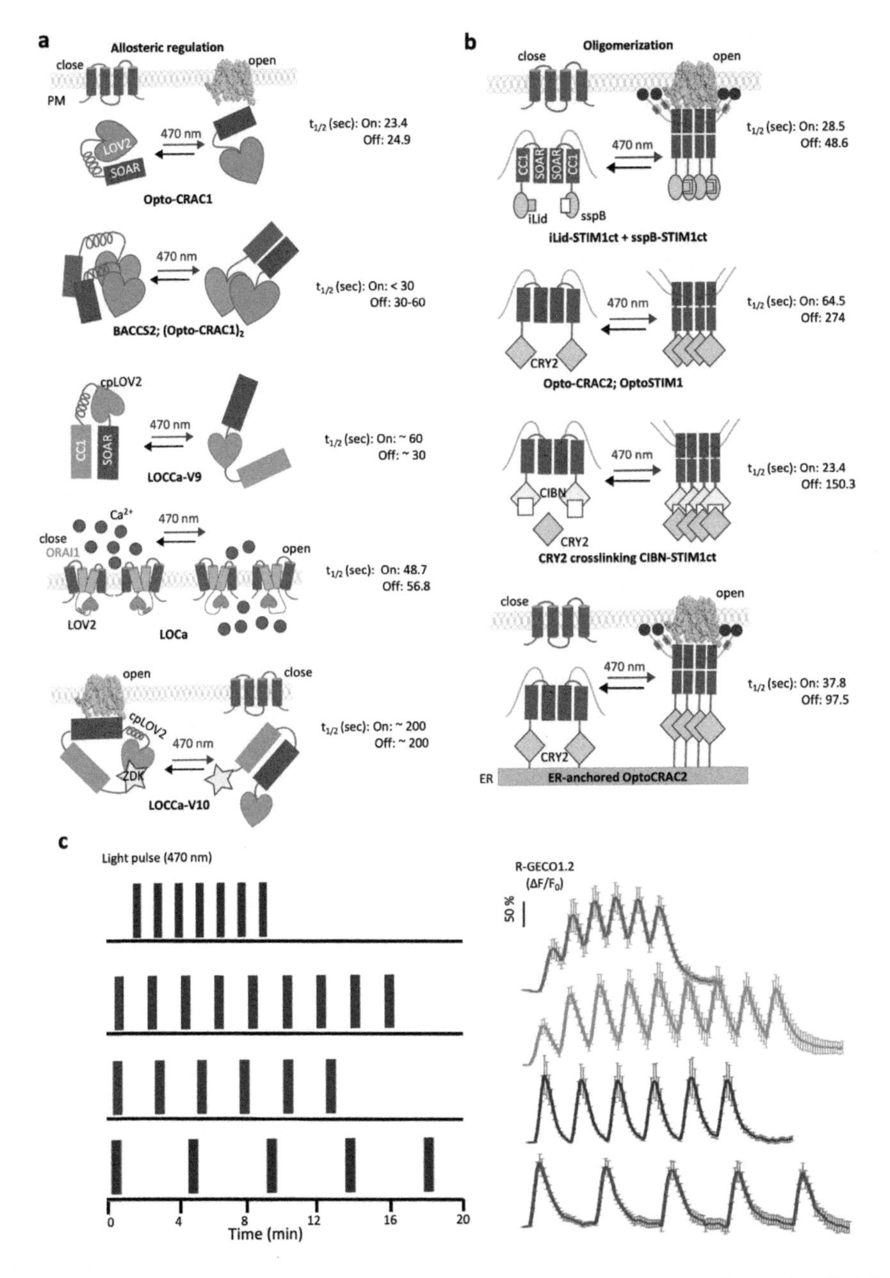

FIGURE 1.3 Various Optogenetic engineering approaches have been applied to the CRAC channel to enable optical control of Ca^{2+} signals with varying kinetic and dynamic properties. (a) LOV2-based allosteric regulation of the CRAC channel (Opto-CRAC1 and BACCS variants: LOV2-based caging SOAR fragments; LOCCa-V9/V10: cpLOV2-based regulation of STIM1ct by inserting different position; LOCa: LOV2 insertion of the intracellular loop of constitutively active ORAI). (b) Oligomerization mediated STIM1 activation and light-inducible gating of the CRAC channel (Opto-CRAC2 or OptoSTIM1 variants). (c) Representative light-tunable Ca^{2+} oscillation patterns generated by HeLa cells expressing Opto-CRAC1. The light pulses were shown on the left.

the best caging in the dark, hence demonstrating a synergistic role of the ID domain in maintaining STIM1ct inactive[24].

CC1-SOAR autoinhibition is essential for keeping STIM1 inactive at the resting state, while SOAR-ORAI1 physical binding is necessary for the gating of the ORAI channel to permit Ca^{2+} influx[7, 15, 70, 84]. It remains challenging to assess the relative binding strengths of SOAR with CC1 or ORAI1 *in cellulo* via conventional biophysical and biochemical methods because of the harsh requirement of biological membranes to keep CC1 and ORAI1 in the correct configuration under physiological conditions[70, 72]. The LOV2-SOAR chimera mentioned previously turned out to be an ideal tool to address this long-standing challenge. LOV2-SOAR chimera

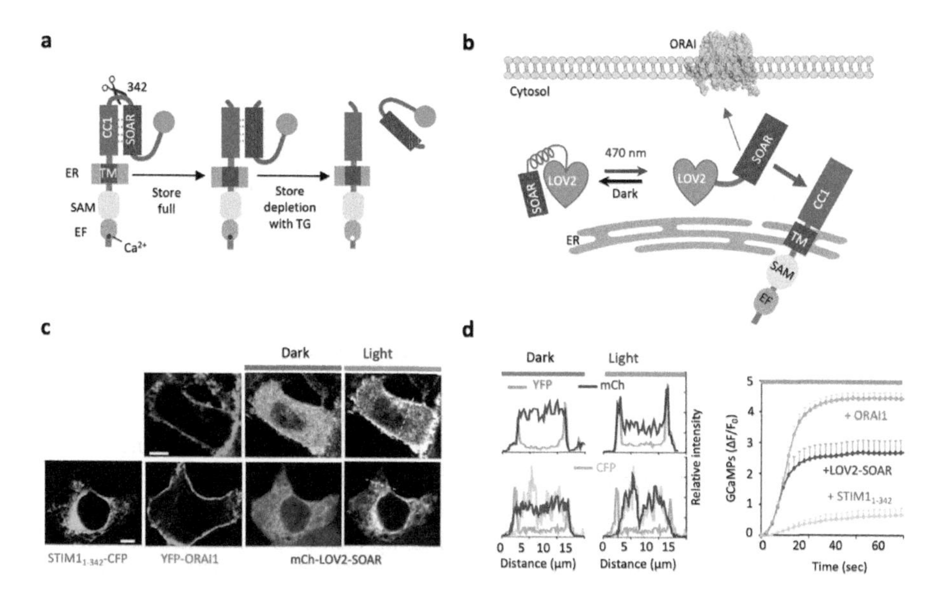

FIGURE 1.4 An engineered LOV2-SOAR chimera used to dissect the "tug of war" between CC1 and ORAI1 during SOCE activation. (a) Schematic of a split STIM1 molecule (at residue 342) to monitor CC1—SOAR interaction in trans. (b) Cartoon illustration of LOV2-SOAR in the dark and lit states. CC1 is replaced by LOV2 to confer photo-controllability. (c-d) Dissection of the relative binding strength of SOAR-CC1 and SOAR-ORAI1 interactions. (c) Top: Light-induced cytosol-to-PM translocation of mCh-LOV2-SOAR (gray) observed in HEK293 cells co-transfected with YFP-ORAI1 (green). Bottom: Confocal images of HEK293 cells coexpressing mCh-LOV2-SOAR (gray), YFP-ORAI (green), and $STIM1_{1-342}$-CFP (cyan). In the dark, mCh-LOV2-SOAR was evenly distributed in the cytosol. Upon photostimulation, mCh-LOV2-SOAR preferentially translocated toward the ER membrane but not to the PM, indicating a stronger interaction between CC1 and SOAR when compared to the SOAR-ORAI1 interaction. (d) The fluorescence intensities (YFP, green; mCh, red; CFP, cyan) across the dashed line were plotted to evaluate the degree of colocalization. (d-right) Light-induced Ca^{2+} response curves (quantified by GCaMP6s) in HEK293 cells transfected with LOV2-SOAR (red), LOV2-SOAR+ORAI1 (green), or LOV2-SOAR+$STIM1_{1-342}$ (blue). CC1-containing $STIM1_{1-342}$ could attenuate Opto-CRAC1 induced Ca^{2+} influx upon photostimulation (blue bar). Scale bar, 5 μm.

was coexpressed with ER-anchored CC1 and PM-localized ORAI at a near 1:1 ratio (Figure 1.4b). LOV2-SOAR showed a smooth cytosolic distribution in the dark and rapidly translocated toward the cytosolic side of the ER membrane, but not the PM, upon photostimulation. This finding indicates a tighter interaction between ER-anchored CC1 and SOAR compared to the SOAR-ORAI1 association (Figure 1.4b-d). In the context of full-length STIM1, it is very likely that other structural elements downstream of the SOAR domain, such as the TRIP and PB motifs discussed subsequently, might lend further support to promote STIM1 translocation toward the cell periphery to win the "tug of war" between CC1 juxta-positioned near the ER membrane and ORAI1 situated in the PM[7]. Collectively, because of its reversibility and light controllability, LOV2-based optogenetic tools provide a unique opportunity to estimate the competitive interactions among signaling domains (CC1-SOAR vs. SOAR-ORAI1, in this case) in living cells[7,24].

1.3.4 Optogenetic Dissection of Protein Self-Oligomerization and PPIs

Ligand-induced protein self-oligomerization and heterotypic PPIs are two themes often seen during cell signaling, including the functional STIM-ORAI coupling during SOCE[64,65,70,85–93]. Following internal Ca^{2+} store depletion, the oligomerization of ER-luminal EF-SAM domain led to the transmission of luminal signals toward the cytosolic side, thereby abolishing CC1-SOAR mediated autoinhibition within STIM1ct to cause CRAC channel activation. Once SOAR is exposed, STIM1 undergoes further oligomerization and subsequently translocates toward the PM. Several techniques, such as Förster resonance energy transfer (FRET), pull-down, and immunoprecipitation, have been applied to examine which domains are required for STIM1 oligomerization and activation[15,70,85,86,90,92], but led to confusion and contradictory conclusions. Compared with conventional biophysical and biochemical approaches, the imaging-based optogenetic assay allows real-time assessment of PPIs at the single-cell level and under more physiologically relevant conditions with simpler procedures. After rounds of engineering and optimizing, *Arabidopsis thaliana* CRY2 has evolved multiple variants with different oligomeric tendencies. For instance, CRY2clust with a C-terminal insertion of a 9-residue peptide shows robust cluster formation (within seconds) after blue light illumination, which can be exploited as an ideal light-triggered clustering tool to study protein-protein interactions by monitoring the co-clustering of candidate bait-prey pairs[7,11,26,28,48,52,81,94,95].

By fusing a domain of interest (termed as "bait") with CRY2clust and coexpressing candidate "prey" proteins within the same cell, a two-color optogenetic clustering assay is well suited to systematically dissect key structural determinants involved in mediating STIM1 oligomerization during SOCE in living cells (Figure 1.5a)[7,11]. If the bait-prey interaction is true, a light-triggered co-clustering is anticipated. Otherwise, one expects to see the clustering of the bait alone, with the prey protein evenly distributed throughout the cytosol. For example, light-induced co-clustering was immediately observed between mCh-CRY2-STIM1$_{233-448}$ (CC1-SOAR) with YFP-STIM1$_{233-685}$ (CC1-SOAR-CT), whereas only the bait protein mCh-CRY2-STIM1$_{233-448}$ formed clustering when coexpressed with YFP-STIM1$_{233-342}$ (CC1) (Figure 1.5b, c, d). By besting various bait-prey combinations with and without the SOAR domain, we concluded that the SOAR domain plays a dominant role in mediating STIMct oligomerization[7]. A similar assay was used to analyze the STIM1 luminal EF-SAM region. It turned out that the SAM domain, but not the EF-hand motif, contributes to the self-oligomerization of the STIM1 luminal domain[7].

Similar CRY2-based optogenetic tools have been further used to address another important problem in STIM1 biology: What is the minimal binding interface between CC1 and SOAR to mediate STIM1 autoinhibition? CC1-SOAR autoinhibition has been confirmed by using recombinant proteins and cell-based assays[67,69,70,87,96,97], and the respective structures of CC1 and SOAR have been determined by nuclear magnetic resonance (NMR) and X-ray crystallography, respectively[67,68,87]. Nonetheless, the molecular determinant underlying CC1-SOAR binding remains elusive. We resorted to optogenetics to map out key regions within CC1 that might mediate intramolecular trapping of SOAR under physiological conditions in living cells[7]. To achieve

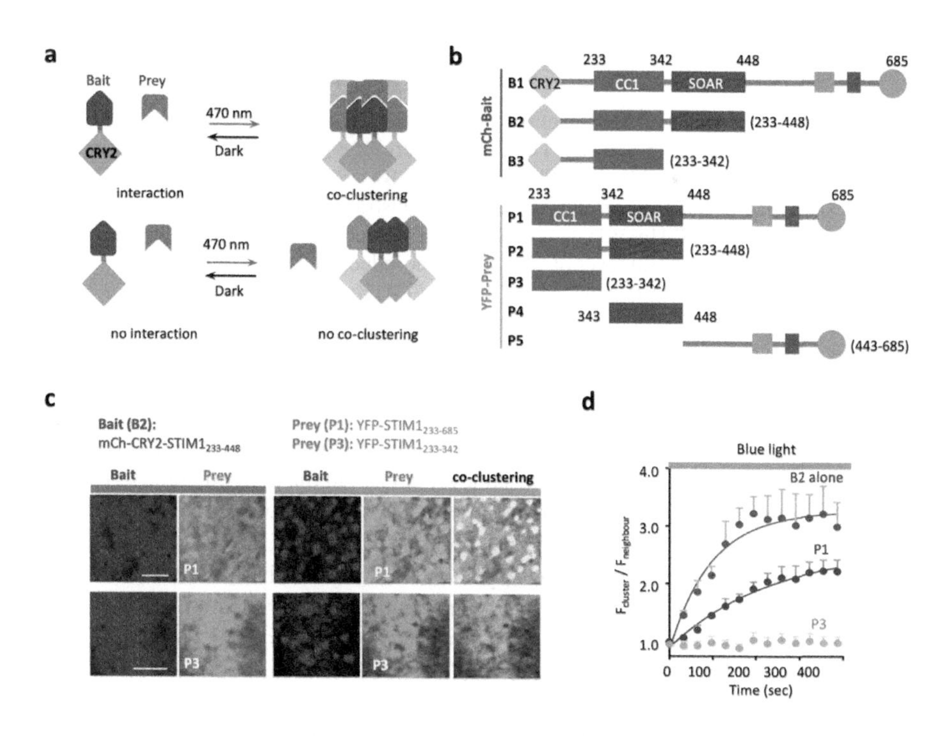

FIGURE 1.5 An optogenetic clustering assay to map domains involved in protein-protein interactions (PPIs). (a) A CRY2-based co-clustering assay for real-time assessment of PPIs in living cells. (b) Summary of mCh-tagged baits and YFP-tagged preys used to map domains critical for STIM1 oligomerization. (c) Confocal images showing HeLa cells expressing the indicated bait-prey pairs (B2-P1 or B2-P3) before and after blue light stimulation. Scale bar, 5 μm (d) Time courses showing the kinetics of light-induced clustering ($F_{cluster}/F_{neighbor}$) of the bait and its co-clustering with P1 (blue), but not with P3 (green), as seen in panel c.

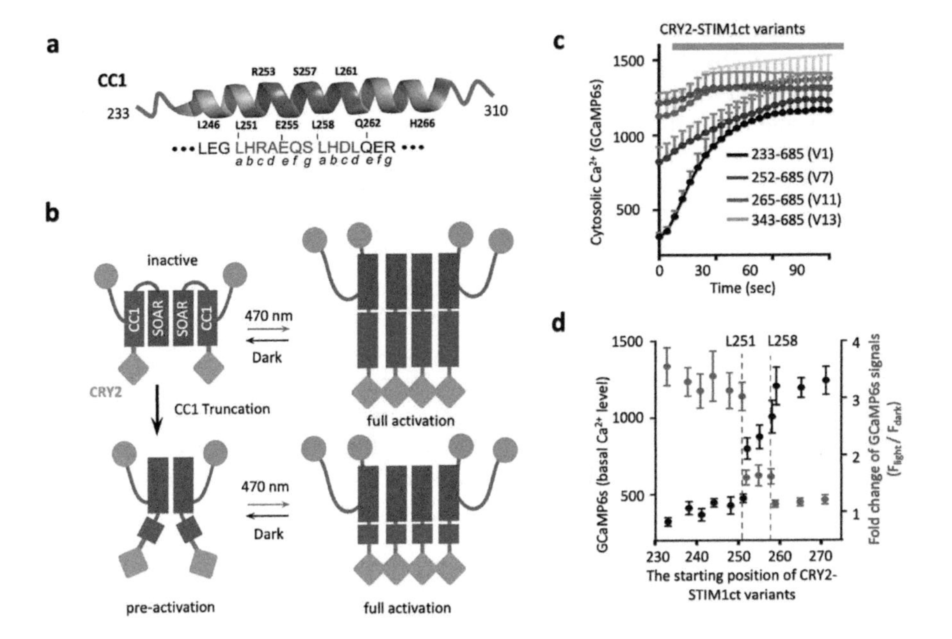

FIGURE 1.6 Optogenetic probing of the binding interface between CC1 and SOAR that mediates STIM1 autoinhibition. (a) The 3D structure of the juxta-ER membrane region of CC1 (selected residues 241–271; PDB entry: 4O9B), with the predicted helical heptad repeats indicated below the cartoon. (b) Cartoon illustrating the anticipated outcomes of the designed CRY2-STIM1ct variants before and after blue light stimulation. (c) GCaMP6s were used to report the basal level of Ca^{2+} and light-induced changes of Ca^{2+} influx. HeLa cells were transfected with the indicated CRY2-STIM1ct constructs. (d) Basal Ca^{2+} levels and fold change in Ca^{2+} signal (light/dark) plotted against the start residues of the CRY2-STIM1ct variants.

this, we fused CRY2 with a set of STIM1ct fragments with CC1 truncated from its N-terminus. We next monitored the basal level of Ca^{2+} in the dark and blue light triggered Ca^{2+} influx as two readouts for evaluating the degree of CC1—SOAR mediated autoinhibition (Figure 1.6b-c). For variants with intact CC1—SOAR interaction (truncation before the position 250), they showed a very low basal level of Ca^{2+} and notable light-induced Ca^{2+} influx. For variants with further truncation of CC1 until position 251, we observed higher basal levels of Ca^{2+} and weaker Ca^{2+} influx induced by light (Figure 1.6d). Based on these results, the N-terminal boundary for SOAR caging was narrowed down to residue L251. Combined with complementary FRET assays to assess interactions between truncated CC1 fragments and SOAR variants, we finally concluded that the polypeptide spanning from L251 to L261 in CC1 is crucial for keeping STIM1ct at an inactive state[7,70], a key finding that was further supported by compelling evidence from recent NMR and single-molecule FRET studies[68,71].

1.3.5 Dissecting STIM1-Organelle Interactions with Optogenetic Approaches

As an ER-anchored protein, STIM1, along with ER network, spreads throughout the cytoplasm to make close contact with multiple organelles and subcellular structures[98]. The S/TxIP motif (TRIP; $STIM1_{642-645}$) and the polybasic domain (PB; $STIM1_{671-685}$) within the STIM1 cytoplasmic tail have been found to interact with EB1 at the plus end of the microtubule and the negatively charged phosphoinositides (PIPs) embedded in the PM, respectively (Figure 1.7a)[37, 74–78]. STIM1 binds to the microtubule-plus-end-tracking protein EB1 via its TRIP sequence, forming comet-like structures to track the microtubule ends and thus remodeling ER structure[74].

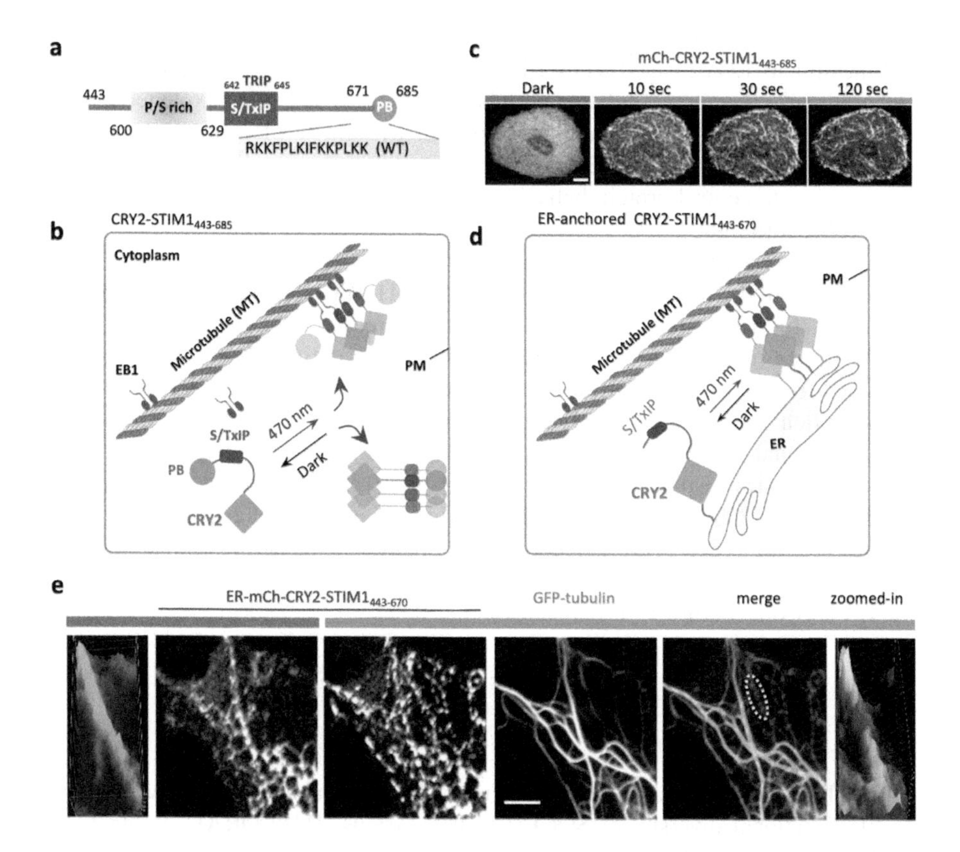

FIGURE 1.7 Optogenetic dissection of the "tug of war" between STIM1-microtubule (MT) and STIM1-PM associations. (a) Domain organization of the STIM1 C-terminal fragment (residues 443–685) containing the S/TxIP EB1-binding motif and a positively charged polybasic tail (PB) that interacts with PM-resident PIPs. (b-c) Confocal images of COS-7 cells expressing mCh-CRY2-$STIM1_{443-685}$ in response to blue light stimulation. Scale bar, 5 μm. (d-e) Light-induced ER-MT interaction in COS-7 cells. Red, ER-anchored mCh-CRY2-$STIM1_{443-670}$; Green, GFP-tubulin. Scale bar, 5 μm.

Chang et al. created an ER-attached chemical inducible oligomerization system by fusing the STIM1 S/TxIP motif with the PB domain to restrict STIM1 targeting to ER-PM junctions, thereby preventing excessive SOCE and ER Ca^{2+} overload[75]. Meanwhile, STIM1 is required for microtubule recruitment and subsequent ER remodeling at the motile side of steering neuronal growth cones[99]. The positively charged PB domain is essential for recruiting activated STIM1 toward ER-PM junctions by engaging PM-embedded PIPs[37, 100].

Optogenetics provides a simple and straightforward method to study protein-organelle interactions. Lian et al. created an optogenetic toolkit, designated as OptoPB, by fusing LOV2 with PB domains derived from GTPases Rit and Rin, STIM1 or myristoylated alanine-rich protein kinase C substrate, MARCKS[79]. OptoPB enables rapid and reversible control of the protein-lipid interaction to enable inducible cytosol-to-PM translocation of a protein of interest. When tethered to the ER membrane, LOV2 caged STIM1-PB (OptoPBer, also termed as LiMETER2 for light-inducible membrane-tethered peripheral ER) could be further exploited to control the assembly of ER-PM membrane contacts by light[37, 79]. By tuning the lengths of linkers, OptoPBer or LiMETER2 can photo-tune the gap distances at ER-PM contact sites to control protein behaviors (e.g., the diffusion of the ORAI1 channel in the PM) with light[79].

A CRY2-fused STIM1 C-terminal fragment (aa 443–685) has also been used to understand the "tug of war" between STIM1-MT and STIM1-PM contacts (Figure 1.7b)[7]. Light-induced CRY2 clustering increased the local avidity to boost STIM1-target interactions. After blue light illumination, the hybrid protein quickly switched from cytosolic distribution to comet-like +TIP tracking, with subsequent translocation toward the PM (Figure 1.7c-d)[7]. When combined with truncation and mutational studies, these optogenetic modules can be used to rapidly identify key residues responsible for the STIM1-organelle interactions[7, 79]. Moreover, ER-anchored CRY2-TIM1ct or LOV2-STIM1ct variants could be applied to dissect STIM1 activation and communications with organelles, including optical manipulations of ER-PM junction formation, ER-MT crosslinking, and the remodeling of the ER network (Figure 1.8)[7, 79].

1.3.6 Optogenetic Dissection of the Functional Differences in STIM Variants

The STIM and ORAI families contain multiple isoforms and splicing variants to enable functional diversity of SOCE[15, 17]. The various STIM-ORAI combinations could generate versatile Ca^{2+} signals with distinct spatiotemporal features and different kinetics. To date, three splicing variants of STIM1 (STIM1L, STIM1β/STIM1A, and STIM1B) and two variants of STIM2 (STIM2.2 and STIM2.3) have been identified with preferred tissue distributions and molecular features[84, 101–103]. Due to highly conserved functional domains and sequence homology, it is always hard to discriminate the functional difference of these isoforms and splice variants.

Optogenetics provides a simple approach to identifying these variants' key structural determinants and dissecting their functional difference. For instance, CRY2

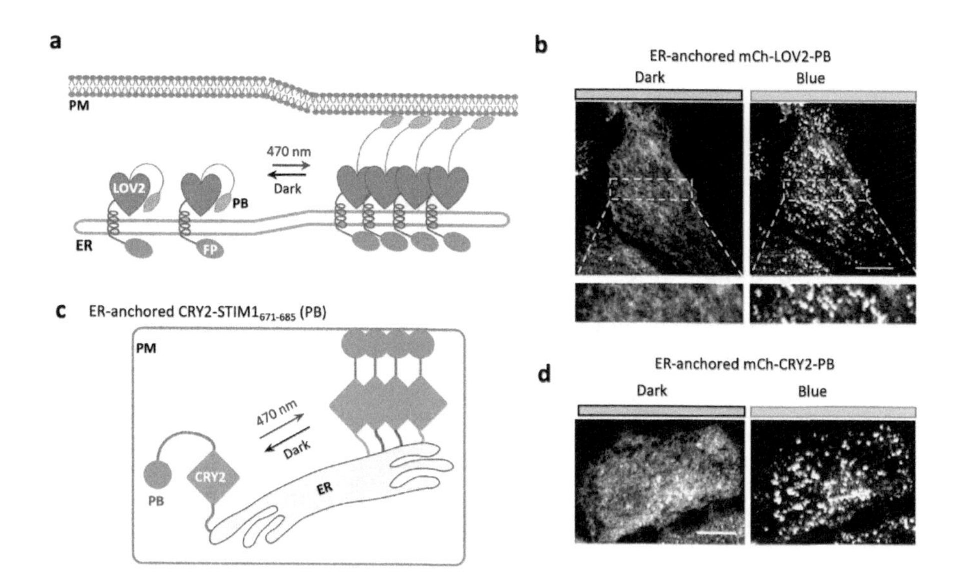

FIGURE 1.8 Engineering STIM1-derived polybasic domain (PB) for optical control of membrane contact site assembly at ER-PM junctions. (a) ER-PM contact sites reversibly labeled with ER-tethered OptoPB (OptoPBer or LiMETER2), which undergoes light-dependent interaction with phosphoinositides embedded in the inner half leaflet of PM. LiMETER can be applied to photo-induce inter-organelle communications to modulate localized Ca^{2+} and lipid signaling. (b) Representative confocal images of ER-tethered mCh-LOV2-PB (or LiMETER2) in the absence or presence of blue light stimulation. Scale bar, 10 μm. (c) Light-induced assembly of ER-plasma membrane contact sites (MCSs) mediated by ER-resident mCh-CRY2-PB. (d) Representative confocal images of ER-tethered mCh-CRY2-PB in the dark (gray bar) and lit conditions (blue bar). Scale bar, 10 μm.

could be fused to various cytosolic fragments derived STIM1 and STIM2 (Figure 1.9a) or variants with hyperactive mutations (E470G and/or L485F) (Figure 1.9b). CRY2-STIM2ct elicited a much weaker increase of intracellular Ca^{2+} when compared to the robust response evoked by CRY2-STIM1ct, confirming STIM2 as a relatively weaker activator of CRAC channels[88]. By contrast, the CRY2-STIM2ct mutants (E470G, L485F, or E470G/L485F) were able to generate pronounced Ca^{2+} responses with rapid kinetics upon light illumination (Figure 1.9c, d), echoing findings made with full-length STIM variants upon store depletion[88]. In another study to compare the performance of a STIM1 isoform with normal STIM1, CRY2-STIM1βct showed a higher basal Ca^{2+} level in the dark but with faster activation kinetics with light, indicating partial pre-activation of STIM1βct[84]. Furthermore, using LOV caged SOAR fragments, Ishii et al. demonstrated that a *Drosophila*-derived LOV2-SOAR chimera (dmBACCS2) failed to activate human ORAI1 (Figure 1.9e)[25]. These features make it possible to develop GECAs capable of versatile control of Ca^{2+} entry with a careful selection of the STIM-ORAI combinations. Clearly, optogenetic approaches can be conveniently adopted to reveal functional differences among isoforms, homologs, and splicing variants of signaling proteins.

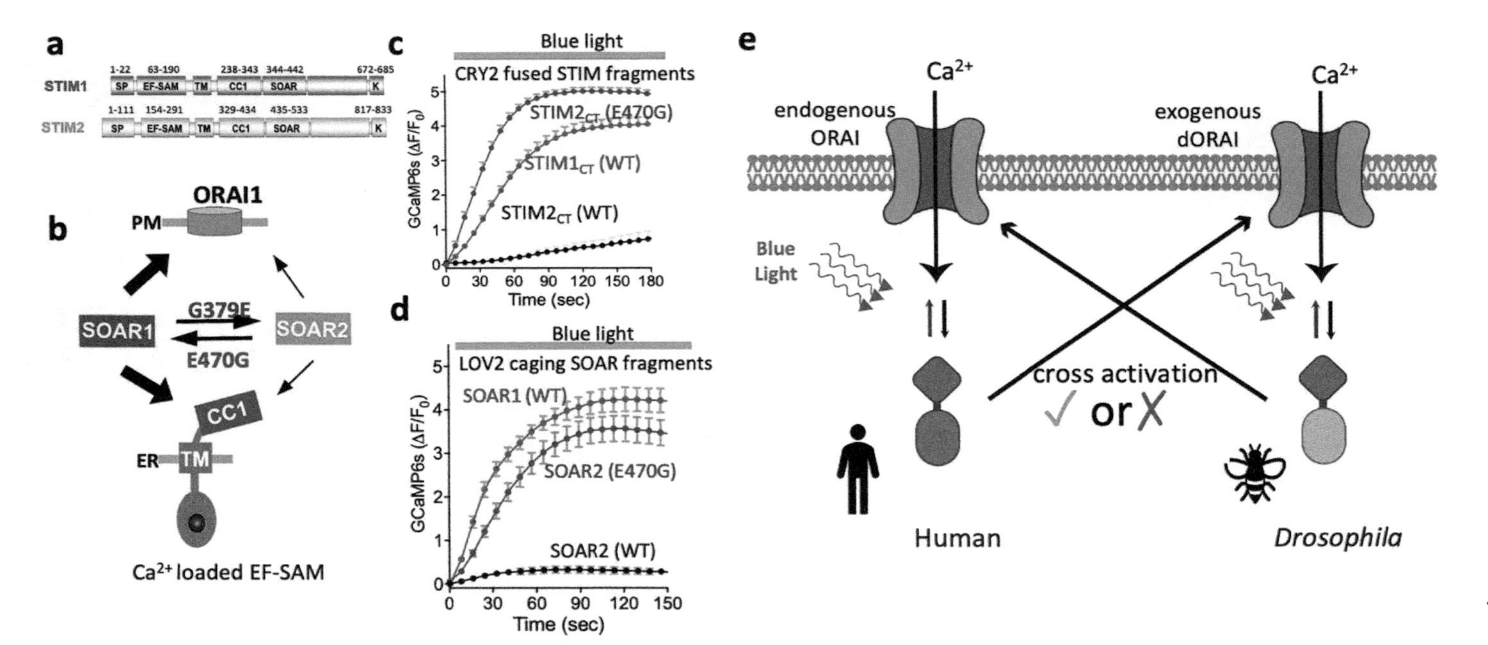

FIGURE 1.9 Optogenetic dissection of the functional difference among STIM variants. (a) Domain architectures of human STIM1 and STIM2. (b) A diagram showing molecular determinants that govern distinct STIM2 activation dynamics. E470 residue in SOAR2-α1 region weakens its interactions with CC1 and ORAI1. E470G mutation makes STIM2 behave more like STIM1. (c-d) Typical Ca^{2+} influx curves in HeLa cells expressing the indicated STIM1/STIM2 variants following light illumination. (e) An optogenetic assay designed to evaluate the cross-activation using *Drosophila* Stim and human ORAI1.

1.4 CONCLUSIONS

By using the two-component STIM-ORAI signaling as a prime example, we have demonstrated the prowess of optogenetic approaches to reveal the mechanistic basis underlying CRAC channel activation. More importantly, these tools can be repurposed to control host cell physiology and benefit disease intervention, as exemplified by our recent development of light-switchable chimeric antigen receptor (LiCAR) T cells[36,104]. CRY2 and LOV2-based tools can be widely adopted to control protein self-oligomerization, protein-protein heterodimerization, and allosteric regulation. Compared to conventional biochemical, pharmacological, and genetic approaches, optogenetic approaches have the advantages of tunable reversibility, high spatiotemporal resolution, and remote controllability during mechanistic dissection of cell signaling. Therefore, it is anticipated that opsin-free optogenetics will likely find broader applications in real-time interrogation of signaling and cell physiology (a sub-field termed "optophysiology") with subcellular resolution ranging from a dozen of milliseconds to minutes.

ACKNOWLEDGMENT

This work was supported by the Welch Foundation (BE-1913–20220331), National Institutes of Health (R01GM112003), and the Cancer Research Institute of Texas (RP210070).

REFERENCES

[1] Fenno, L., Yizhar, O., and Deisseroth, K. (2011) The development and application of optogenetics, *Annu Rev Neurosci 34*, 389–412.

[2] Tan, P., He, L., Huang, Y., and Zhou, Y. (2022) Optophysiology: Illuminating cell physiology with optogenetics, *Physiol Rev 102*, 1263–1325.

[3] Toettcher, J. E., Voigt, C. A., Weiner, O. D., and Lim, W. A. (2011) The promise of optogenetics in cell biology: Interrogating molecular circuits in space and time, *Nat Methods 8*, 35–38.

[4] Deisseroth, K. (2011) Optogenetics, *Nat Methods 8*, 26–29.

[5] Tischer, D., and Weiner, O. D. (2014) Illuminating cell signalling with optogenetic tools, *Nat Rev Mol Cell Biol 15*, 551–558.

[6] Oh, T. J., Fan, H., Skeeters, S. S., and Zhang, K. (2021) Steering molecular activity with optogenetics: Recent advances and perspectives, *Adv Biol (Weinh) 5*, e2000180.

[7] Ma, G., He, L., Liu, S., Xie, J., Huang, Z., Jing, J., Lee, Y. T., Wang, R., Luo, H., Han, W., Huang, Y., and Zhou, Y. (2020) Optogenetic engineering to probe the molecular choreography of STIM1-mediated cell signaling, *Nat Commun 11*, 1039.

[8] Kamps, D., Koch, J., Juma, V. O., Campillo-Funollet, E., Graessl, M., Banerjee, S., Mazel, T., Chen, X., Wu, Y. W., Portet, S., Madzvamuse, A., Nalbant, P., and Dehmelt, L. (2020) Optogenetic tuning reveals RHO amplification-dependent dynamics of a cell contraction signal network, *Cell Rep 33*, 108467.

[9] Bagci, H., Sriskandarajah, N., Robert, A., Boulais, J., Elkholi, I. E., Tran, V., Lin, Z. Y., Thibault, M. P., Dube, N., Faubert, D., Hipfner, D. R., Gingras, A. C., and Cote, J. F. (2020) Mapping the proximity interaction network of the Rho-family GTPases reveals signalling pathways and regulatory mechanisms, *Nat Cell Biol 22*, 120–134.

[10] de Beco, S., Vaidziulyte, K., Manzi, J., Dalier, F., di Federico, F., Cornilleau, G., Dahan, M., and Coppey, M. (2018) Optogenetic dissection of Rac1 and Cdc42 gradient shaping, *Nat Commun 9*, 4816.
[11] Taslimi, A., Vrana, J. D., Chen, D., Borinskaya, S., Mayer, B. J., Kennedy, M. J., and Tucker, C. L. (2014) An optimized optogenetic clustering tool for probing protein interaction and function, *Nat Commun 5*, 4925.
[12] Zhang, K., Duan, L., Ong, Q., Lin, Z., Varman, P. M., Sung, K., and Cui, B. (2014) Light-mediated kinetic control reveals the temporal effect of the Raf/MEK/ERK pathway in PC12 cell neurite outgrowth, *PLoS One 9*, e92917.
[13] Krishnamurthy, V. V., Fu, J., Oh, T. J., Khamo, J., Yang, J., and Zhang, K. (2020) A generalizable optogenetic strategy to regulate receptor tyrosine kinases during vertebrate embryonic development, *J Mol Biol 432*, 3149–3158.
[14] Parekh, A. B., and Putney, J. W., Jr. (2005) Store-operated calcium channels, *Physiol Rev 85*, 757–810.
[15] Prakriya, M., and Lewis, R. S. (2015) Store-operated calcium channels, *Physiol Rev 95*, 1383–1436.
[16] Hogan, P. G., Lewis, R. S., and Rao, A. (2010) Molecular basis of calcium signaling in lymphocytes: STIM and ORAI, *Annu Rev Immunol 28*, 491–533.
[17] Soboloff, J., Rothberg, B. S., Madesh, M., and Gill, D. L. (2012) STIM proteins: Dynamic calcium signal transducers, *Nat Rev Mol Cell Biol 13*, 549–565.
[18] Vaeth, M., Kahlfuss, S., and Feske, S. (2020) CRAC channels and calcium signaling in T cell-mediated immunity, *Trends Immunol 41*, 878–901.
[19] Lewis, R. S. (2020) Store-operated calcium channels: From function to structure and back again, *Cold Spring Harb Perspect Biol 12*.
[20] Woll, K. A., and Van Petegem, F. (2022) Calcium-release channels: Structure and function of IP3 receptors and ryanodine receptors, *Physiol Rev 102*, 209–268.
[21] Nguyen, N. T., Ma, G., Zhou, Y., and Jing, J. (2020) Optogenetic approaches to control Ca(2+)-modulated physiological processes, *Curr Opin Physiol 17*, 187–196.
[22] Ma, G., Wen, S., He, L., Huang, Y., Wang, Y., and Zhou, Y. (2017) Optogenetic toolkit for precise control of calcium signaling, *Cell Calcium 64*, 36–46.
[23] Ma, G., and Zhou, Y. (2020) A STIMulating journey into optogenetic engineering, *Cell Calcium 88*, 102197.
[24] He, L., Zhang, Y., Ma, G., Tan, P., Li, Z., Zang, S., Wu, X., Jing, J., Fang, S., Zhou, L., Wang, Y., Huang, Y., Hogan, P. G., Han, G., and Zhou, Y. (2015) Near-infrared photoactivatable control of Ca(2+) signaling and optogenetic immunomodulation, *Elife 4*.
[25] Ishii, T., Sato, K., Kakumoto, T., Miura, S., Touhara, K., Takeuchi, S., and Nakata, T. (2015) Light generation of intracellular Ca(2+) signals by a genetically encoded protein BACCS, *Nat Commun 6*, 8021.
[26] Kyung, T., Lee, S., Kim, J. E., Cho, T., Park, H., Jeong, Y. M., Kim, D., Shin, A., Kim, S., Baek, J., Kim, J., Kim, N. Y., Woo, D., Chae, S., Kim, C. H., Shin, H. S., Han, Y. M., Kim, D., and Heo, W. D. (2015) Optogenetic control of endogenous Ca(2+) channels in vivo, *Nat Biotechnol 33*, 1092–1096.
[27] Hannanta-Anan, P., and Chow, B. Y. (2016) Optogenetic control of calcium oscillation waveform defines NFAT as an integrator of calcium load, *Cell Syst 2*, 283–288.
[28] Kim, S., Kyung, T., Chung, J. H., Kim, N., Keum, S., Lee, J., Park, H., Kim, H. M., Lee, S., Shin, H. S., and Heo, W. D. (2020) Non-invasive optical control of endogenous Ca(2+) channels in awake mice, *Nat Commun 11*, 210.
[29] He, L., Tan, P., Zhu, L., Huang, K., Nguyen, N. T., Wang, R., Guo, L., Li, L., Yang, Y. H., Huang, Z. X., Huang, Y., Han, G., Wang, J. F., and Zhou, Y. B. (2021) Circularly permuted LOV2 as a modular photoswitch for optogenetic engineering, *Nature Chemical Biology 17*, 915–923.

[30] He, L., Wang, L., Zeng, H., Tan, P., Ma, G., Zheng, S., Li, Y., Sun, L., Dou, F., Siwko, S., Huang, Y., Wang, Y., and Zhou, Y. (2021) Engineering of a bona fide light-operated calcium channel, *Nat Commun 12*, 164.

[31] Wu, Y. I., Frey, D., Lungu, O. I., Jaehrig, A., Schlichting, I., Kuhlman, B., and Hahn, K. M. (2009) A genetically encoded photoactivatable Rac controls the motility of living cells, *Nature 461*, 104–108.

[32] Christie, J. M. (2007) Phototropin blue-light receptors, *Annu Rev Plant Biol 58*, 21–45.

[33] Peter, E., Dick, B., and Baeurle, S. A. (2010) Mechanism of signal transduction of the LOV2-Jalpha photosensor from Avena sativa, *Nat Commun 1*, 122.

[34] Zoltowski, B. D., Schwerdtfeger, C., Widom, J., Loros, J. J., Bilwes, A. M., Dunlap, J. C., and Crane, B. R. (2007) Conformational switching in the fungal light sensor Vivid, *Science 316*, 1054–1057.

[35] Motta-Mena, L. B., Reade, A., Mallory, M. J., Glantz, S., Weiner, O. D., Lynch, K. W., and Gardner, K. H. (2014) An optogenetic gene expression system with rapid activation and deactivation kinetics, *Nat Chem Biol 10*, 196–202.

[36] He, L., Tan, P., Zhu, L., Huang, K., Nguyen, N. T., Wang, R., Guo, L., Li, L., Yang, Y., Huang, Z., Huang, Y., Han, G., Wang, J., and Zhou, Y. (2021) Circularly permuted LOV2 as a modular photoswitch for optogenetic engineering, *Nat Chem Biol 17*, 915–923.

[37] Jing, J., He, L., Sun, A., Quintana, A., Ding, Y., Ma, G., Tan, P., Liang, X., Zheng, X., Chen, L., Shi, X., Zhang, S. L., Zhong, L., Huang, Y., Dong, M. Q., Walker, C. L., Hogan, P. G., Wang, Y., and Zhou, Y. (2015) Proteomic mapping of ER-PM junctions identifies STIMATE as a regulator of Ca(2)(+) influx, *Nat Cell Biol 17*, 1339–1347.

[38] Mathony, J., and Niopek, D. (2021) Enlightening allostery: Designing switchable proteins by photoreceptor fusion, *Adv Biol (Weinh) 5*, e2000181.

[39] Pudasaini, A., El-Arab, K. K., and Zoltowski, B. D. (2015) LOV-based optogenetic devices: Light-driven modules to impart photoregulated control of cellular signaling, *Front Mol Biosci 2*, 18.

[40] Spiltoir, J. I., and Tucker, C. L. (2019) Photodimerization systems for regulating protein-protein interactions with light, *Curr Opin Struct Biol 57*, 1–8.

[41] Guntas, G., Hallett, R. A., Zimmerman, S. P., Williams, T., Yumerefendi, H., Bear, J. E., and Kuhlman, B. (2015) Engineering an improved light-induced dimer (iLID) for controlling the localization and activity of signaling proteins, *Proc Natl Acad Sci U S A 112*, 112–117.

[42] Zoltowski, B. D., and Crane, B. R. (2008) Light activation of the LOV protein vivid generates a rapidly exchanging dimer, *Biochemistry 47*, 7012–7019.

[43] Ni, M., Tepperman, J. M., and Quail, P. H. (1999) Binding of phytochrome B to its nuclear signalling partner PIF3 is reversibly induced by light, *Nature 400*, 781–784.

[44] Liu, H., Yu, X., Li, K., Klejnot, J., Yang, H., Lisiero, D., and Lin, C. (2008) Photoexcited CRY2 interacts with CIB1 to regulate transcription and floral initiation in Arabidopsis, *Science 322*, 1535–1539.

[45] Park, S. Y., and Tame, J. R. H. (2017) Seeing the light with BLUF proteins, *Biophys Rev 9*, 169–176.

[46] Palayam, M., Ganapathy, J., Guercio, A. M., Tal, L., Deck, S. L., and Shabek, N. (2021) Structural insights into photoactivation of plant Cryptochrome-2, *Commun Biol 4*, 28.

[47] Ma, L., Wang, X., Guan, Z., Wang, L., Wang, Y., Zheng, L., Gong, Z., Shen, C., Wang, J., Zhang, D., Liu, Z., and Yin, P. (2020) Structural insights into BIC-mediated inactivation of Arabidopsis cryptochrome 2, *Nat Struct Mol Biol 27*, 472–479.

[48] Bugaj, L. J., Choksi, A. T., Mesuda, C. K., Kane, R. S., and Schaffer, D. V. (2013) Optogenetic protein clustering and signaling activation in mammalian cells, *Nat Methods 10*, 249–252.

[49] Chang, K. Y., Woo, D., Jung, H., Lee, S., Kim, S., Won, J., Kyung, T., Park, H., Kim, N., Yang, H. W., Park, J. Y., Hwang, E. M., Kim, D., and Heo, W. D. (2014) Light-inducible receptor tyrosine kinases that regulate neurotrophin signalling, *Nat Commun 5*, 4057.
[50] Kim, N., Kim, J. M., Lee, M., Kim, C. Y., Chang, K. Y., and Heo, W. D. (2014) Spatiotemporal control of fibroblast growth factor receptor signals by blue light, *Chem Biol 21*, 903–912.
[51] Bugaj, L. J., Spelke, D. P., Mesuda, C. K., Varedi, M., Kane, R. S., and Schaffer, D. V. (2015) Regulation of endogenous transmembrane receptors through optogenetic Cry2 clustering, *Nat Commun 6*, 6898.
[52] Park, H., Kim, N. Y., Lee, S., Kim, N., Kim, J., and Heo, W. D. (2017) Optogenetic protein clustering through fluorescent protein tagging and extension of CRY2, *Nat Commun 8*, 30.
[53] Tan, P., He, L., and Zhou, Y. (2021) Engineering supramolecular organizing centers for optogenetic control of innate immune responses, *Adv Biol (Weinh) 5*, e2000147.
[54] He, L., Huang, Z., Huang, K., Chen, R., Nguyen, N. T., Wang, R., Cai, X., Huang, Z., Siwko, S., Walker, J. R., Han, G., Zhou, Y., and Jing, J. (2021) Optogenetic control of non-apoptotic cell death, *Adv Sci (Weinh) 8*, 2100424.
[55] Zhou, X. X., Chung, H. K., Lam, A. J., and Lin, M. Z. (2012) Optical control of protein activity by fluorescent protein domains, *Science 338*, 810–814.
[56] Zhao, E. M., Suek, N., Wilson, M. Z., Dine, E., Pannucci, N. L., Gitai, Z., Avalos, J. L., and Toettcher, J. E. (2019) Light-based control of metabolic flux through assembly of synthetic organelles, *Nat Chem Biol 15*, 589–597.
[57] Stone, O. J., Pankow, N., Liu, B., Sharma, V. P., Eddy, R. J., Wang, H., Putz, A. T., Teets, F. D., Kuhlman, B., Condeelis, J. S., and Hahn, K. M. (2019) Optogenetic control of cofilin and alphaTAT in living cells using Z-lock, *Nat Chem Biol 15*, 1183–1190.
[58] Wang, H., Vilela, M., Winkler, A., Tarnawski, M., Schlichting, I., Yumerefendi, H., Kuhlman, B., Liu, R., Danuser, G., and Hahn, K. M. (2016) LOVTRAP: An optogenetic system for photoinduced protein dissociation, *Nat Methods 13*, 755–758.
[59] Zhou, X. X., Fan, L. Z., Li, P., Shen, K., and Lin, M. Z. (2017) Optical control of cell signaling by single-chain photoswitchable kinases, *Science 355*, 836–842.
[60] De Smet, F., Christopoulos, A., and Carmeliet, P. (2014) Allosteric targeting of receptor tyrosine kinases, *Nat Biotechnol 32*, 1113–1120.
[61] Nussinov, R., Tsai, C. J., and Liu, J. (2014) Principles of allosteric interactions in cell signaling, *J Am Chem Soc 136*, 17692–17701.
[62] Manoilov, K. Y., Verkhusha, V. V., and Shcherbakova, D. M. (2021) A guide to the optogenetic regulation of endogenous molecules, *Nat Methods 18*, 1027–1037.
[63] Dagliyan, O., Dokholyan, N. V., and Hahn, K. M. (2019) Engineering proteins for allosteric control by light or ligands, *Nat Protoc 14*, 1863–1883.
[64] Stathopulos, P. B., Li, G. Y., Plevin, M. J., Ames, J. B., and Ikura, M. (2006) Stored Ca2+ depletion-induced oligomerization of stromal interaction molecule 1 (STIM1) via the EF-SAM region: An initiation mechanism for capacitive Ca2+ entry, *J Biol Chem 281*, 35855–35862.
[65] Stathopulos, P. B., Zheng, L., Li, G. Y., Plevin, M. J., and Ikura, M. (2008) Structural and mechanistic insights into STIM1-mediated initiation of store-operated calcium entry, *Cell 135*, 110–122.
[66] Gudlur, A., Zeraik, A. E., Hirve, N., Rajanikanth, V., Bobkov, A. A., Ma, G., Zheng, S., Wang, Y., Zhou, Y., Komives, E. A., and Hogan, P. G. (2018) Calcium sensing by the STIM1 ER-luminal domain, *Nat Commun 9*, 4536.
[67] Yang, X., Jin, H., Cai, X., Li, S., and Shen, Y. (2012) Structural and mechanistic insights into the activation of Stromal interaction molecule 1 (STIM1), *Proc Natl Acad Sci U S A 109*, 5657–5662.

[68] Rathner, P., Fahrner, M., Cerofolini, L., Grabmayr, H., Horvath, F., Krobath, H., Gupta, A., Ravera, E., Fragai, M., Bechmann, M., Renger, T., Luchinat, C., Romanin, C., and Muller, N. (2021) Interhelical interactions within the STIM1 CC1 domain modulate CRAC channel activation, *Nat Chem Biol 17*, 196–204.
[69] Kim, J. Y., and Muallem, S. (2011) Unlocking SOAR releases STIM, *EMBO J 30*, 1673–1675.
[70] Ma, G., Wei, M., He, L., Liu, C., Wu, B., Zhang, S. L., Jing, J., Liang, X., Senes, A., Tan, P., Li, S., Sun, A., Bi, Y., Zhong, L., Si, H., Shen, Y., Li, M., Lee, M. S., Zhou, W., Wang, J., Wang, Y., and Zhou, Y. (2015) Inside-out Ca(2+) signalling prompted by STIM1 conformational switch, *Nat Commun 6*, 7826.
[71] van Dorp, S., Qiu, R., Choi, U. B., Wu, M. M., Yen, M., Kirmiz, M., Brunger, A. T., and Lewis, R. S. (2021) Conformational dynamics of auto-inhibition in the ER calcium sensor STIM1, *Elife 10*.
[72] Zhou, Y., Meraner, P., Kwon, H. T., Machnes, D., Oh-hora, M., Zimmer, J., Huang, Y., Stura, A., Rao, A., and Hogan, P. G. (2010) STIM1 gates the store-operated calcium channel ORAI1 in vitro, *Nat Struct Mol Biol 17*, 112–116.
[73] Gudlur, A., Quintana, A., Zhou, Y., Hirve, N., Mahapatra, S., and Hogan, P. G. (2014) STIM1 triggers a gating rearrangement at the extracellular mouth of the ORAI1 channel, *Nat Commun 5*, 5164.
[74] Grigoriev, I., Gouveia, S. M., van der Vaart, B., Demmers, J., Smyth, J. T., Honnappa, S., Splinter, D., Steinmetz, M. O., Putney, J. W., Jr., Hoogenraad, C. C., and Akhmanova, A. (2008) STIM1 is a MT-plus-end-tracking protein involved in remodeling of the ER, *Curr Biol 18*, 177–182.
[75] Chang, C. L., Chen, Y. J., Quintanilla, C. G., Hsieh, T. S., and Liou, J. (2018) EB1 binding restricts STIM1 translocation to ER-PM junctions and regulates store-operated Ca(2+) entry, *J Cell Biol 217*, 2047–2058.
[76] Luik, R. M., Wu, M. M., Buchanan, J., and Lewis, R. S. (2006) The elementary unit of store-operated Ca2+ entry: local activation of CRAC channels by STIM1 at ER-plasma membrane junctions, *J Cell Biol 174*, 815–825.
[77] Varnai, P., Toth, B., Toth, D. J., Hunyady, L., and Balla, T. (2007) Visualization and manipulation of plasma membrane-endoplasmic reticulum contact sites indicates the presence of additional molecular components within the STIM1-Orai1 Complex, *J Biol Chem 282*, 29678–29690.
[78] Balla, T. (2018) Ca(2+) and lipid signals hold hands at endoplasmic reticulum-plasma membrane contact sites, *J Physiol 596*, 2709–2716.
[79] He, L., Jing, J., Zhu, L., Tan, P., Ma, G., Zhang, Q., Nguyen, N. T., Wang, J., Zhou, Y., and Huang, Y. (2017) Optical control of membrane tethering and interorganellar communication at nanoscales, *Chem Sci 8*, 5275–5281.
[80] Pham, E., Mills, E., and Truong, K. (2011) A synthetic photoactivated protein to generate local or global Ca(2+) signals, *Chem Biol 18*, 880–890.
[81] Bohineust, A., Garcia, Z., Corre, B., Lemaitre, F., and Bousso, P. (2020) Optogenetic manipulation of calcium signals in single T cells in vivo, *Nat Commun 11*, 1143.
[82] Yang, X., Ma, G., Zheng, S., Qin, X., Li, X., Du, L., Wang, Y., Zhou, Y., and Li, M. (2020) Optical control of CRAC channels using photoswitchable azopyrazoles, *J Am Chem Soc 142*, 9460–9470.
[83] Nesin, V., Wiley, G., Kousi, M., Ong, E. C., Lehmann, T., Nicholl, D. J., Suri, M., Shahrizaila, N., Katsanis, N., Gaffney, P. M., Wierenga, K. J., and Tsiokas, L. (2014) Activating mutations in STIM1 and ORAI1 cause overlapping syndromes of tubular myopathy and congenital miosis, *Proc Natl Acad Sci U S A 111*, 4197–4202.
[84] Xie, J., Ma, G., Zhou, L., He, L., Zhang, Z., Tan, P., Huang, Z., Fang, S., Wang, T., Lee, Y. T., Wen, S., Siwko, S., Wang, L., Liu, J., Du, Y., Zhang, N., Liu, X., Han, L., Huang,

Y., Wang, R., Wang, Y., Zhou, Y., and Han, W. (2022) Identification of a STIM1 splicing variant that promotes glioblastoma growth, *Adv Sci (Weinh)*, e2103940.

[85] Liou, J., Fivaz, M., Inoue, T., and Meyer, T. (2007) Live-cell imaging reveals sequential oligomerization and local plasma membrane targeting of stromal interaction molecule 1 after Ca2+ store depletion, *Proc Natl Acad Sci U S A 104*, 9301–9306.

[86] Luik, R. M., Wang, B., Prakriya, M., Wu, M. M., and Lewis, R. S. (2008) Oligomerization of STIM1 couples ER calcium depletion to CRAC channel activation, *Nature 454*, 538–542.

[87] Cui, B., Yang, X., Li, S., Lin, Z., Wang, Z., Dong, C., and Shen, Y. (2013) The inhibitory helix controls the intramolecular conformational switching of the C-terminus of STIM1, *PLoS One 8*, e74735.

[88] Zheng, S., Ma, G., He, L., Zhang, T., Li, J., Yuan, X., Nguyen, N. T., Huang, Y., Zhang, X., Gao, P., Nwokonko, R., Gill, D. L., Dong, H., Zhou, Y., and Wang, Y. (2018) Identification of molecular determinants that govern distinct STIM2 activation dynamics, *PLoS Biol 16*, e2006898.

[89] Park, C. Y., Hoover, P. J., Mullins, F. M., Bachhawat, P., Covington, E. D., Raunser, S., Walz, T., Garcia, K. C., Dolmetsch, R. E., and Lewis, R. S. (2009) STIM1 clusters and activates CRAC channels via direct binding of a cytosolic domain to Orai1, *Cell 136*, 876–890.

[90] Covington, E. D., Wu, M. M., and Lewis, R. S. (2010) Essential role for the CRAC activation domain in store-dependent oligomerization of STIM1, *Mol Biol Cell 21*, 1897–1907.

[91] Lee, S. K., Lee, M. H., Jeong, S. J., Qin, X., Lee, A. R., Park, H., and Park, C. Y. (2020) The inactivation domain of STIM1 acts through intramolecular binding to the coiled-coil domain in the resting state, *J Cell Sci 133*.

[92] Maus, M., Jairaman, A., Stathopulos, P. B., Muik, M., Fahrner, M., Weidinger, C., Benson, M., Fuchs, S., Ehl, S., Romanin, C., Ikura, M., Prakriya, M., and Feske, S. (2015) Missense mutation in immunodeficient patients shows the multifunctional roles of coiled-coil domain 3 (CC3) in STIM1 activation, *Proc Natl Acad Sci U S A 112*, 6206–6211.

[93] Li, Z., Lu, J., Xu, P., Xie, X., Chen, L., and Xu, T. (2007) Mapping the interacting domains of STIM1 and Orai1 in Ca2+ release-activated Ca2+ channel activation, *J Biol Chem 282*, 29448–29456.

[94] Gil, A. A., Carrasco-Lopez, C., Zhu, L., Zhao, E. M., Ravindran, P. T., Wilson, M. Z., Goglia, A. G., Avalos, J. L., and Toettcher, J. E. (2020) Optogenetic control of protein binding using light-switchable nanobodies, *Nat Commun 11*, 4044.

[95] Kennedy, M. J., Hughes, R. M., Peteya, L. A., Schwartz, J. W., Ehlers, M. D., and Tucker, C. L. (2010) Rapid blue-light-mediated induction of protein interactions in living cells, *Nat Methods 7*, 973–975.

[96] Korzeniowski, M. K., Manjarres, I. M., Varnai, P., and Balla, T. (2010) Activation of STIM1-Orai1 involves an intramolecular switching mechanism, *Sci Signal 3*, ra82.

[97] Fahrner, M., Muik, M., Schindl, R., Butorac, C., Stathopulos, P., Zheng, L., Jardin, I., Ikura, M., and Romanin, C. (2014) A coiled-coil clamp controls both conformation and clustering of stromal interaction molecule 1 (STIM1), *J Biol Chem 289*, 33231–33244.

[98] English, A. R., and Voeltz, G. K. (2013) Endoplasmic reticulum structure and interconnections with other organelles, *Cold Spring Harb Perspect Biol 5*, a013227.

[99] Pavez, M., Thompson, A. C., Arnott, H. J., Mitchell, C. B., D'Atri, I., Don, E. K., Chilton, J. K., Scott, E. K., Lin, J. Y., Young, K. M., Gasperini, R. J., and Foa, L. (2019) STIM1 is required for remodeling of the endoplasmic reticulum and microtubule cytoskeleton in steering growth cones, *J Neurosci 39*, 5095–5114.

[100] Wu, M. M., Buchanan, J., Luik, R. M., and Lewis, R. S. (2006) Ca2+ store depletion causes STIM1 to accumulate in ER regions closely associated with the plasma membrane, *J Cell Biol 174*, 803–813.
[101] Rosado, J. A., Diez, R., Smani, T., and Jardin, I. (2015) STIM and orai1 variants in store-operated calcium entry, *Front Pharmacol 6*, 325.
[102] Ramesh, G., Jarzembowski, L., Schwarz, Y., Poth, V., Konrad, M., Knapp, M. L., Schwar, G., Lauer, A. A., Grimm, M. O. W., Alansary, D., Bruns, D., and Niemeyer, B. A. (2021) A short isoform of STIM1 confers frequency-dependent synaptic enhancement, *Cell Rep 34*, 108844.
[103] Knapp, M. L., Alansary, D., Poth, V., Forderer, K., Sommer, F., Zimmer, D., Schwarz, Y., Kunzel, N., Kless, A., Machaca, K., Helms, V., Muhlhaus, T., Schroda, M., Lis, A., and Niemeyer, B. A. (2021) A longer isoform of Stim1 is a negative SOCE regulator but increases cAMP-modulated NFAT signaling, *EMBO Rep*, e53135.
[104] Nguyen, N. T., Huang, K., Zeng, H., Jing, J., Wang, R., Fang, S., Chen, J., Liu, X., Huang, Z., You, M. J., Rao, A., Huang, Y., Han, G., and Zhou, Y. (2021) Nano-optogenetic engineering of CAR T cells for precision immunotherapy with enhanced safety, *Nat Nanotechnol 16*, 1424–1434.

2 High-Throughput Engineering of a Light-Activatable Ca^{2+} Channel

Lian He, Liuqing Wang, and Youjun Wang

CONTENTS

2.1 Introduction 25
2.2 Methods 27
2.2.1 Generation of an ORAI1 Expression Vector 27
2.2.2 Making a Library of LOV2-ORAI1 Hybrid Constructs 27
2.2.3 Cell-Based Screening with Ca^{2+} Imaging 28
2.2.4 Generation of a LOCa Library by Error-Prone PCR 30
2.2.5 NFAT Translocation-Based High-Content Imaging 32
2.2.6 Confirmation of Top Hits with NFAT-Dependent Luciferase Expression 32
2.3 Summary 33
Acknowledgments 33
References 34

2.1 INTRODUCTION

To enable optical control of Ca^{2+} mobilization and the activation of downstream signaling cascades in mammalian cells, multiple optogenetic engineering approaches have been applied to modulate the Ca^{2+} handling machinery [1–4]. Among them, the microbial opsin-based channelrhodopsin-2 (ChR2) is a single-component optogenetic device that can induce Ca^{2+} influx upon blue light stimulation. However, in spite of efforts to improve their Ca^{2+} permeability [5], this family of light-gated ion channels also conduct other cation ions and generally lack high selectivity for Ca^{2+} [6, 7]. To overcome this hurdle, store-operated Ca^{2+} release-activated Ca^{2+} (CRAC) channel that has a strict Ca^{2+} selectivity [8] was chosen as a target to develop optogenetic Ca^{2+} signaling toolkits by multiple groups [1]. A prototypical CRAC channel consists of the endoplasmic reticulum (ER)-resident Ca^{2+} sensor protein, stromal interaction molecule 1 (STIM1), and ORAI1, a four-pass plasma membrane (PM) protein that assembles into a hexamer to form the pore subunit of CRAC channel [9–11]. Under physiological conditions, STIM1 senses ER Ca^{2+} depletion, and undergoes conformational changes to interact with ORAI1 and open the ORAI channel. Subsequently, Ca^{2+} influx through the ORAI pore could further activate a downstream Ca^{2+} responsive transcriptional factor, the nuclear factor of activated-T cells

DOI: 10.1201/b22823-2

(NFAT), which subsequently translocates from the cytoplasm into the nucleus to turn on gene expression [11–13]. Researchers have engineered plant-derived photosensory domains into STIM1 to generate light-operated Ca^{2+} channel actuators. Because of the high spatiotemporal resolution of light, low cytotoxicity, fast kinetics, and superior reversibility, light-induced Ca^{2+}/NFAT signaling has been widely used for *ex vivo* manipulation of cytokine production in immune cells [14, 15]. These STIM1-based genetically encoded Ca^{2+} actuators (GECAs) operate in two ways: (1) light-dependent conformational change of LOV2 (light-oxygen-voltage) domain from *Avena sativa* phototropin 1 to uncage an ORAI-activating fragment (SOAR or CAD) derived from STIM1; or (2) oligomerization of the PHR (photolyase-homology) region of *Arabidopsis thaliana* cryptochrome 2 (CRY2) to drive STIM1 cytoplasmic domain (STIM1ct) activation [14–18]. However, STIM1-based GECAs require endogenous ORAI expression and cannot activate cells with no or low ORAI levels [19]. In addition, crosstalk with other STIM1-associated targets cannot be ignored in STIM1-based optogenetic Ca^{2+} actuators [20–22]. To overcome these caveats associated with ChR2 and STIM1-based designs, we generated a blue light-operated Ca^{2+} (LOCa) channel by directly installing LOV2 domain into ORAI1 to generate a single-component compact light-operated Ca^{2+} channel [19] (Figure 2.1).

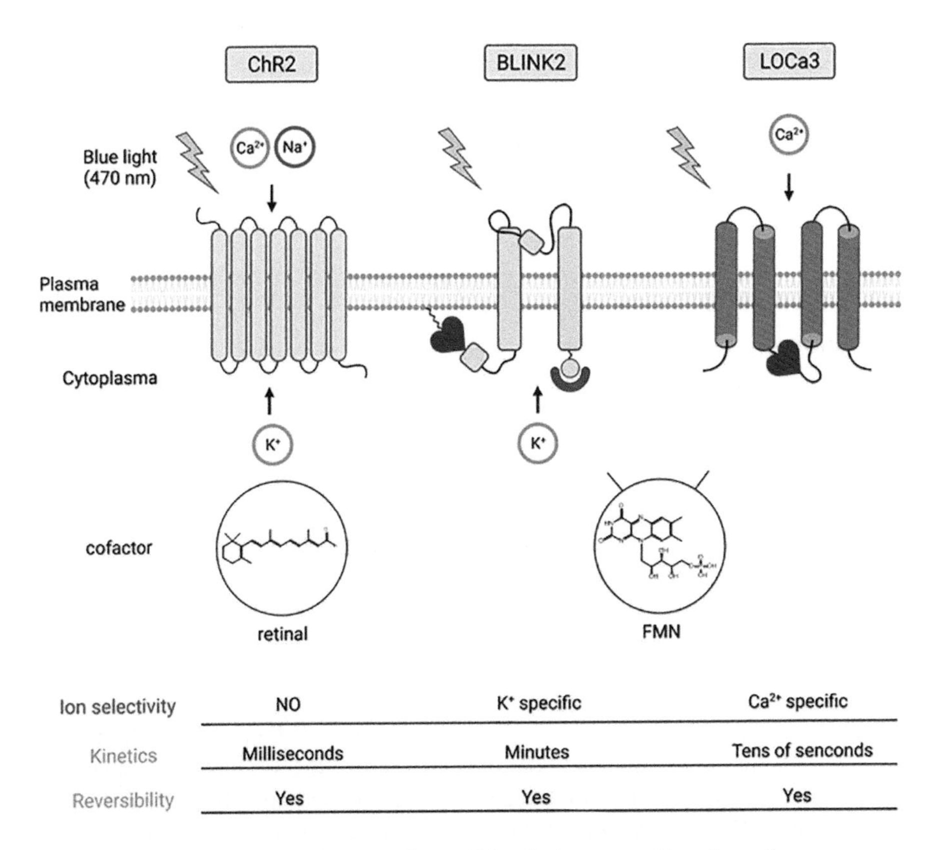

FIGURE 2.1 Comparison of commonly used blue light-operated ion channels.

In this chapter, we outline our procedures for generating, optimizing, and validating LOCa constructs (Figure 2.2). We described a fluorescence imaging-based high-throughput protocol to effectively screen and improve LOCa. Although we used ORAI1 as an engineering example, our general approach can be readily extended to photo-manipulate other types of ion channels with high precision.

2.2 METHODS

2.2.1 Generation of an ORAI1 Expression Vector

A regular mammalian expression vector was used for ORAI1 expression, and a red fluorescent protein, mKate2, was used here as an indicator for ORAI1 expression levels. To avoid interfering with the normal subcellular distribution of ORAI1 and subsequent protein engineering, we did not directly fuse mKate with ORAI1. Instead, mKate2 and ORAI1 were linked with a 2A self-cleaving peptide (P2A, ATNFSLLKQAGDVEENPGP) derived from porcine teschovirus-1, which allows equal expression of both mKate2 and ORAI1 in the same cell. The mKate2 fragment with P2A was first cloned into a pcDNA3.1(+) vector with a CMV promotor by using the standard enzyme digestion and cloning method. Subsequently, cDNA encoding human ORAI1 was amplified by using KOD Hot Start DNA polymerase (EMD Millipore, MA, USA) and inserted into the vector to make pcDNA3.1(+)-mKate2-P2A-ORAI1. To avoid possible interference from those extra residues that were left after the self-cleavage of the p2A peptide, we chose to fuse P2A at N-terminus of ORAI1, so that the ORAI1 C-end is intact for proper function.

2.2.2 Making a Library of LOV2-ORAI1 Hybrid Constructs

LOV2 domain from *Avena sativa* phototropin 1 is a commonly used photosensor that can change its conformation when exposed to blue light at 450–470 nm. LOV2 contains a per-arnt-sim (PAS) core domain and C-terminal Jα helix, with flavin mononucleotide (FMN) cofactor. After absorbing blue light, a covalent bond is formed between residue C450 and FMN, which leads to the unfolding of the Jα helix to release the C-terminal effector protein [23]. LOV2 domain has been used as an allosteric switch to regulate protein-protein interaction [24], degradation [25], and Ca^{2+} or GTPase signaling [14, 15, 26]. However, there are very limited examples of applying LOV2 to control membrane-embedded proteins. Prior to the development of LOCa, the only example was BLINK, which was designed by fusing the LOV2 domain to the N-terminus of a viral potassium channel Kcv to control K^+ flux with blue light [27, 28]. Compared with Kcv that only contains two transmembrane (TM) segments, ORAI1 is a more challenging target for optogenetic engineering since it contains four TM regions and assembles as a hexamer in the PM. Considering that LOV2 can be inserted into a host protein and used as an allosteric switch to optically-induce conformational change [29], we tested a few insertion sites in ORAI1. A computer-based method was used to predict which loop region is suitable for LOV2 fusion or insertion [30]. To reduce the energy cost for the allosteric switch that leads to ORAI1 activation, we found that it is better to start with constitutively active

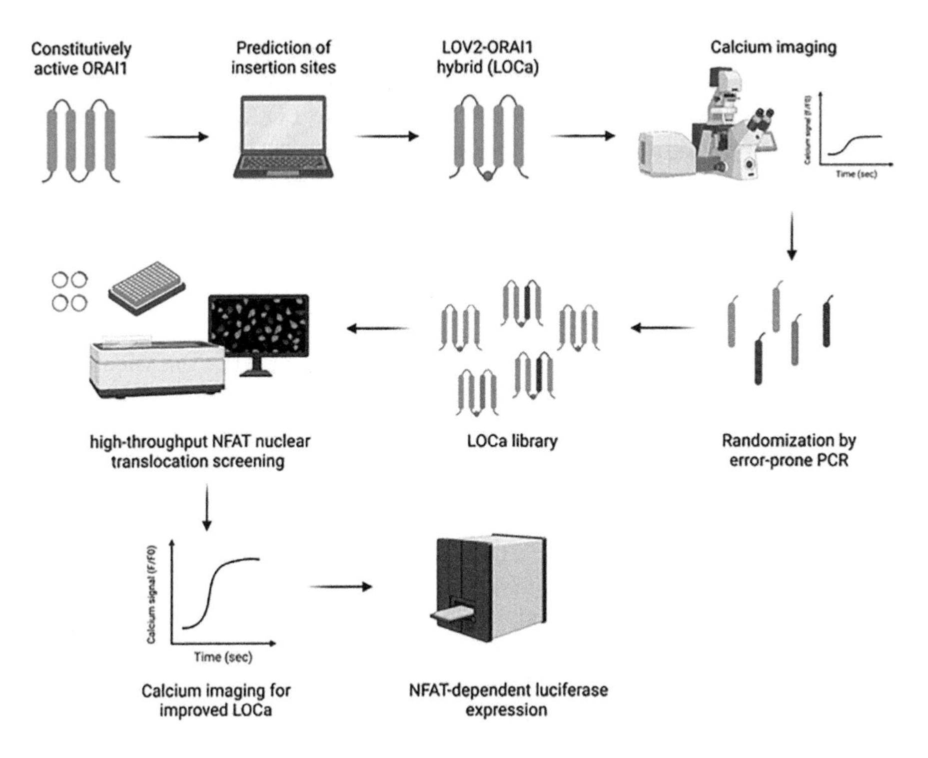

FIGURE 2.2 Outline of the experimental procedures for LOCa engineering and screening.

mutants (H134A or P245T) instead of the wild-type (WT) ORAI1. We used the NEBuilder HiFi DNA Assembly strategy (NEBuilder HiFi DNA Assembly Master Mix, New England Biolabs Inc.) to make LOV2-ORAI1 hybrid (LOCa) constructs, as this method is easy to use, highly efficient, and not dependent on restriction sites. BamHI and BspEI restriction sites were often introduced to encode GS/SG as flanking linkers when LOV2 was inserted into ORAI1.

2.2.3 Cell-Based Screening with Ca^{2+} Imaging

HeLa and human embryonic kidney (HEK) 293 cells are commonly used cell lines because of their high transfection efficiency. HeLa cells were preferred in this cell-based Ca^{2+} imaging since they can adhere tightly to the bottom of the culture dishes and maintain a good morphology during time-lapse confocal imaging. HeLa cells were cultured in Dulbecco's modified Eagle's medium supplemented with 10% heat-inactivated fetal bovine serum and 1% penicillin-streptomycin.

To monitor LOCa-mediated Ca^{2+} influx, both small molecule Ca^{2+} dyes and genetically encoded Ca^{2+} indicators (GECIs) can be used. Small-molecule Ca^{2+} dyes such as Fluo-4 AM require pre-incubation, are more expensive, and show toxicity to cells during chronic experiments. By contrast, GECIs like GCaMP are cost-effective, bear negligible toxicity, and are suitable for long-term imaging [27]. Depending on the purpose of the planned screening projects, one needs to consider four major factors with regard to the choice of appropriate GECIs: basal fluorescence, sensitivity, speed, and dynamics. The latest GCaMP family GECI, jGCaMP8, showed fast kinetics and high sensitivity for both *in vivo* and *in vitro* Ca^{2+} imaging but had a relatively smaller dynamic range in our hand [28]. In our experiments, we used jGCaMP7c because of its high sensitivity and large dynamic changes in fluorescence upon Ca^{2+} binding [31].

Day 1: Cell seeding

1. Carefully discard the medium from the culture plate and rinse adherent cells gently with pre-warmed PBS.
2. Add trypsin-EDTA to the cell monolayer and put the cell plate back into an incubator at 37°C for 2–3 mins to detach cells, then resuspend cells with pre-warmed medium and transfer them to a 15 ml conical bottom centrifuge tube.
3. Centrifuge at 800 × g for 5 mins, discard the supernatant and suspend cell pellet with 0.5–1 ml fresh medium.
4. Count cells using an automated cell counter, and then seed cells at a 1×10^5 cells/ml density for 500 µL into a 4-chamber 35 mm glass-bottom dish (Cellvis, cat #D35C4–20–0-N). If using a 35 mm glass-bottom dish with one well (MatTek, cat #P35G-1.5–14-C), add 2 ml suspended HeLa cells to each well and gently shake the plate to distribute the seeded cells evenly. We typically perform Ca^{2+} imaging with a 40 × or 60 × oil lens using a confocal microscope and therefore prefer to use a glass-bottom imaging dish without further steps to transfer cells to coverslips for imaging.

Day 2: Transfection

The next day, the cells are expected to reach >70% confluency and are ready for transfection. A total of 200 ng plasmids encoding LOCa variants with different LOV2 insertion sites and ORAI1-activating mutations (one variant at a time), together with jGCaMP7c, were co-transfected with Lipofectamine 3000 (Thermo Fisher Scientific, MA, USA) following the manufacturer's protocol. After incubating the plasmids-lipofectamine mixture at room temperature for 15 min, add the DNA-lipid complex to the cells. Keep the plate back in an incubator at 37°C with 5% CO_2. The cells were maintained in the dark to reduce potential exposure to room light.

Day 3: Ca^{2+} imaging

1. To monitor the blue light-induced Ca^{2+} changes in LOCa over-expressing cells, a Nikon W1 Yokogawa spinning disk confocal microscope with a photo-stimulation micro-scanner was used. An environmentally controlled incubator is most desirable to maintain the proper temperature at 37°C and humidity for long-term live-cell imaging.

2. Find mKate2-positive transfected cells under a 40 × or 60 × microscope objective lens and turn on the "Perfect Focus System (PFS)" module. Do not pre-expose cells to blue light or the EGFP excitation channel to avoid pre-activation of the optogenetic tools.
3. Record time-lapse imaging for both green (jGCaMP7c) and red (mKate2-P2A-LOCa) channels every 2 sec for 3 mins. In most situations, the 488 nm laser for the green channel is strong enough to photo-activate LOV2, so an external blue light source stimulation is not necessary. Repeat the previous steps for 3–5 fields to record enough number of cells.
4. To quantify jGCaMP7c intensity, define the cells of interest by the region-of-interest (ROI) toolbox in the Nikon NIS-Elements software. Next, measure the intensity change by using the "Time Measurement" tool. Calculate the jGCaMP7c fold-change F_{max}/F_0 for mKate2 positive cells, using the change of mKate2-negative cells as a negative control. F_{max}, maximal fluorescent signal with light stimulation; F_0, the basal fluorescent signal in the dark. The LOCa variant with the largest F_{max}/F_0, or the largest light-induced Ca^{2+} influx, will be further optimized.

2.2.4 Generation of a LOCa Library by Error-Prone PCR

In the initial round of screening, there were almost no light-induced intracellular Ca^{2+} changes for ORAI1 constructs with direct LOV2 fusion. By contrast, light-induced Ca^{2+} entry from an ORAI1-P245T hybrid with LOV2 insertion in its intracellular loop (between TM2 and TM3) was observed. This prompted us to carry out further improvements. Since TM3-TM4 helix coupling is important for STIM1-induced ORAI1 activation [32], we envisioned that introducing random mutations across the TM3-TM4 by error-prone PCR might evolve LOCa with better performance. From previous Ca^{2+} imaging, the screened constitutively active mutant P245T is located at the TM4, so we focused on randomly mutating the TM3 and the second extracellular loop region (residues 160–233). The GeneMorph II Random Mutagenesis Kit (Agilent, Santa Clara, CA, USA) was used to generate the mutant library. The mutation frequency is related to the initial target amount and ampliation thermal cycles. For the ORAI fragment in a size of about 200 bp, 1–2 mutations needed to be induced, with the mutation frequency of 5 mutations/kb. According to the manufacturer's instruction, 100–500 ng initial target amount was used in a total 50 µl reaction volume with 33 PCR cycles. The purified PCR mutate library was digested with BspEI and XmaI and inserted into mKate2-P2A-LOCa to replace the corresponding ORAI fragment to make the LOCa variant library. BspEI was introduced into LOCa by LOV2 insertion, and XmaI is an endogenous site in the second extracellular loop of ORAI1. Next, to achieve the highest transformation efficiency of the ligation products, we recommend using electrotransformation with commercially available high-efficiency competent cells. Pick single colonies (300–500, more if needed) and culture them in LB medium overnight, and then isolate the corresponding plasmid DNA with a plasmid miniprep kit (Figure 2.3).

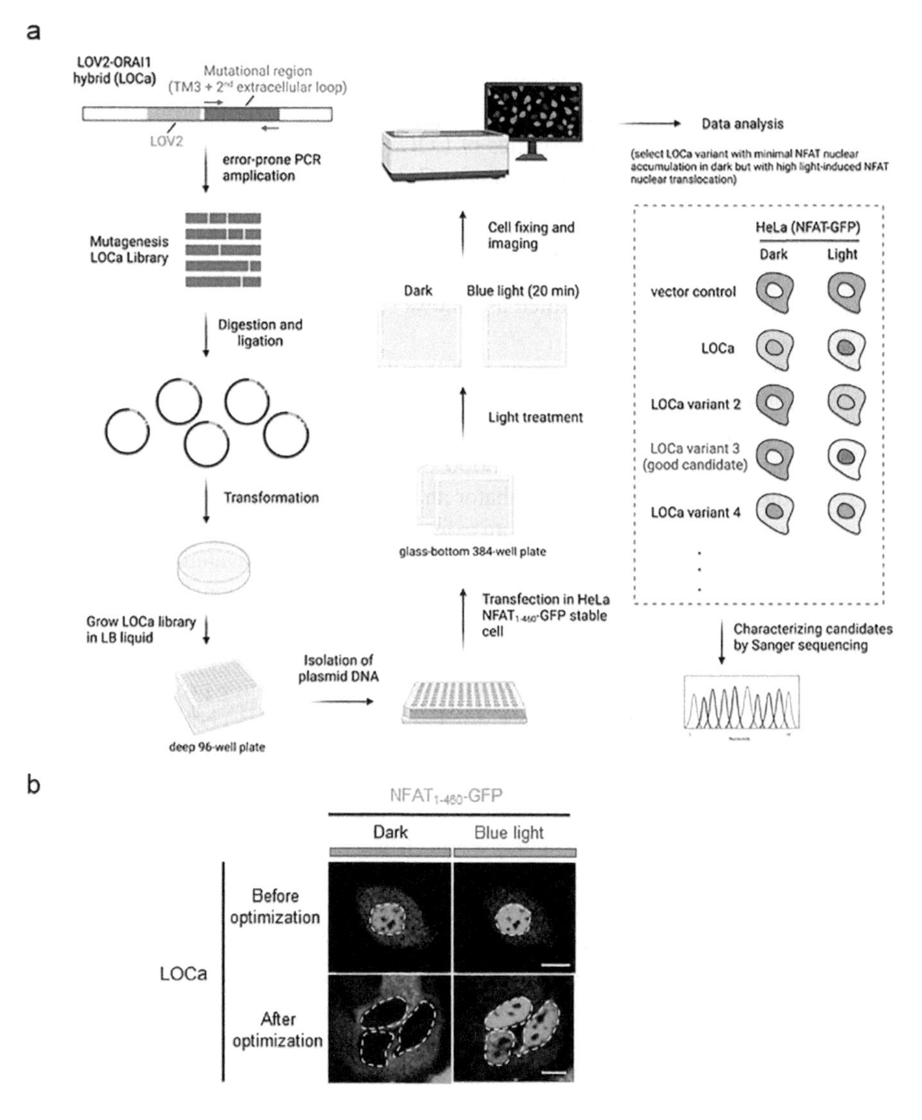

FIGURE 2.3 LOCa optimization and characterization. (a) Error-prone PCR was used to amplify the TM3 and 2nd extracellular region of LOCa. The resulting LOCa variant library was transfected into HeLa $NFAT_{1-460}$-GFP stable cell line and treated with or without blue light for next round of screening. Last, characterize the optimized LOCa by Sanger sequencing. (b) Validation of optimized LOCa variant in $NFAT_{1-460}$-GFP stable cells. Cell expressing Optimized LOCa showed low dark activation and high light-induced NFAT nuclear accumulation. Scale bar, 10 μm.

2.2.5 NFAT Translocation-Based High-Content Imaging

The jGCaMP7 Ca^{2+} sensor is an intensiometric indicator that only indicates light-induced Ca^{2+} changes mediated by LOCa variants, but cannot estimate the dark activity. To estimate the dark activities of LOCa variants, we resorted to the downstream transcription factor NFAT, using its Ca^{2+}-triggered cytosol-to-nuclear translocation as a readout to reflect the Ca^{2+} signals in both the dark and lit states. To reduce the heterogeneous protein level of $NFAT_{1-460}$-GFP from transient transfection, we generated HeLa cells stably expressing $NFAT_{1-460}$-GFP for high-throughput screening. The procedure is detailed as follows (Figure 2.3):

1. Pipette 25 μL of cell suspension (2000 cells) into each well of a black 384-well glass-bottom plate. After 18–24 hours, transfect LOCa mutant library (single variant for each well) into HeLa ($NFAT_{1-460}$-GFP) cells with 25 ng plasmids/well by Lipofectamine 3000. Set up four repeats for each construct.
2. Put the plate back into the incubator and keep cells cultured in the dark for 24 hours. Illuminate cells with an external LED blue light (470 nm, 40 μW/mm^2, ThorLabs, Inc.) for 20 min to trigger the light-activated Ca^{2+}/NFAT signaling. At the same time, a duplicate plate was continuously kept in the dark to measure the dark activity (or pre-activation in the dark).
3. Cells were washed with PBS, fixed with 4% paraformaldehyde (PFA), and their nuclei stained with DAPI using a standard cell staining protocol. Fixed cells can be stored in PBS at 4°C for 1–2 days (in the dark) before imaging.
4. Acquire high-throughput imaging by using an IN Cell Analyzer 6000 Cell Imaging System (GE-Healthcare Life Sciences) with automation for mKate2, GFP, and DAPI channels, with 2 fields for each well. A 10× or higher objective lens is enough for NFAT-based screening.
5. High-content imaging data were analyzed with Pipeline Pilot by calculating the ratio of nuclear over cytosolic fluorescence intensity of NFAT-GFP. Plot the ratios for each construct at both dark and light-treated conditions. The ideal candidates should exhibit minimal basal activity but with high light-induced NFAT nuclear translocation. The candidates will be Sanger sequenced to validate the mutations. Typical images of LOCa-induced Ca^{2+} signals are shown in Figure 2.4. The reversibility of LOCa-operated Ca^{2+} influx can be monitored by red GECIs, and external LED blue light was applied as the trigger.

2.2.6 Confirmation of Top Hits with NFAT-Dependent Luciferase Expression

The next step is to examine downstream NFAT-dependent gene expression by using luciferase as a convenient readout. HEK293T cells were placed in a white 96-well cell culture microplate at a density of 5×10^4 cells/well. 100 ng mKate2-P2A-LOCa or mKate2-P2A-LOCa(E106A) was co-transfected with 30 ng pGL4.30[luc2P/NFAT-RE/Hygro] (NFAT RE-Luc, Promega, Madison, WI, USA) with three biological replicates for each group. mKate2-P2A-LOCa(E106A) is the ORAI pore-dead

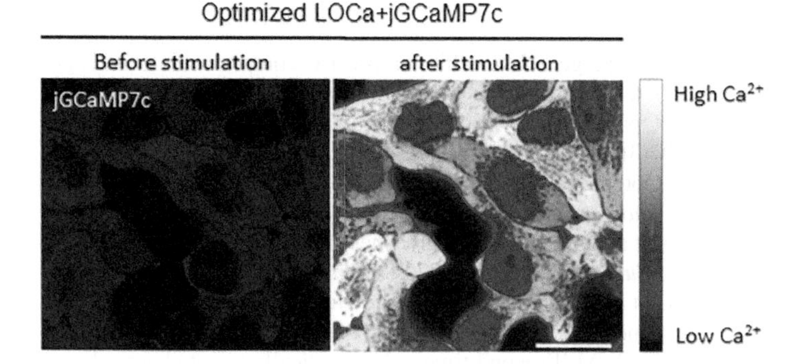

FIGURE 2.4 Blue light-induced cytosolic Ca^{2+} increase in HeLa cells mediated by optimized LOCa. Hela cells were transiently expressed with an optimized LOCa and the green-color Ca^{2+} indicator jGCaMP7c. Typical pictures show pseudocolor images for jGCaMP7c before and after stimulation with blue light (2 min, 40 μW/mm^2). Warm colors reflect high cytosolic Ca^{2+} concentrations. Scale bar, 10 μm.

negative control. Sixteen hours later, cells were treated with or without photostimulation (470 nm, 40 μW/mm^2, 1 min light every 10 min) for 10 hours. At the same time, phorbol myristate acetate (PMA) was added to activate the co-stimulatory pathway. The Bright-Glo Luciferase Assay System from Promega (catalog #: E2610) was prepared as instructed and directly added to cell cultures, with luminescence signals measured by a Cytation 5 Cell Imaging Multi-Mode Reader. Cells expressing LOCa are anticipated to exhibit robust luminescence signals, while those with pore-dead LOCa-E106A display negligible changes in luminescence, thus confirming that the light-induced Ca^{2+} signals are indeed specifically mediated by LOCa.

2.3 SUMMARY

In conclusion, we have provided a guideline for optogenetic engineering of Ca^{2+} channels by using ORAI1 as a prime example. By using the LOV2 domain as an allosteric switch and insertion into intracellular loop regions of ORAI1, we have generated light-controllable LOV2-ORAI1 chimeras with high spatial-temporal resolution. After randomized mutations through error-prone PCR and high-throughput imaging screening, the best performing candidate, termed as LOCa, has been evolved with the lowest dark activities and highest Ca^{2+} response upon blue light stimulation. Collectively, this high-throughput screening platform allows us not only to develop a single-component light-gate ORAI1 channel with high Ca^{2+} selectivity but also can be widely adopted for engineering other proteins or channels.

ACKNOWLEDGMENTS

This work was supported by the National Natural Science foundation of China (91954205 to Y.W.), and the Ministry of Science and Technology of China (grant 2019YFA0802104 to Y.W.).

REFERENCES

1. Ma, G., et al., Optogenetic toolkit for precise control of calcium signaling. *Cell Calcium*, 2017. **64**: pp. 36–46.
2. Tan, P., et al., Optophysiology: Illuminating cell physiology with optogenetics. *Physiol Rev*, 2022. **102**(3): pp. 1263–1325.
3. Tan, P., et al., Optogenetic immunomodulation: Shedding light on antitumor immunity. *Trends Biotechnol*, 2017. **35**(3): pp. 215–226.
4. Nguyen, N.T., et al., Optogenetic approaches to control Ca(2+)-modulated physiological processes. *Curr Opin Physiol*, 2020. **17**: pp. 187–196.
5. Kleinlogel, S., et al., Ultra light-sensitive and fast neuronal activation with the Ca(2)+-permeable channelrhodopsin CatCh. *Nat Neurosci*, 2011. **14**(4): pp. 513–518.
6. Boyden, E.S., et al., Millisecond-timescale, genetically targeted optical control of neural activity. *Nat Neurosci*, 2005. **8**(9): pp. 1263–1268.
7. Zhang, F., et al., Multimodal fast optical interrogation of neural circuitry. *Nature*, 2007. **446**(7136): pp. 633–639.
8. Prakriya, M. and R.S. Lewis, Store-operated calcium channels. *Physiol Rev*, 2015. **95**(4): pp. 1383–1436.
9. Soboloff, J., et al., STIM proteins: Dynamic calcium signal transducers. *Nat Rev Mol Cell Biol*, 2012. **13**(9): pp. 549–565.
10. Putney, J.W., et al., The functions of store-operated calcium channels. *Biochim Biophys Acta Mol Cell Res*, 2017. **1864**(6): pp. 900–906.
11. Hogan, P.G. and A. Rao, Store-operated calcium entry: Mechanisms and modulation. *Biochem Biophys Res Commun*, 2015. **460**(1): pp. 40–49.
12. Feske, S., H. Wulff, and E.Y. Skolnik, Ion channels in innate and adaptive immunity. *Annu Rev Immunol*, 2015. **33**: pp. 291–353.
13. Hogan, P.G., R.S. Lewis, and A. Rao, Molecular basis of calcium signaling in lymphocytes: STIM and ORAI. *Annu Rev Immunol*, 2010. **28**: pp. 491–533.
14. Ma, G., et al., Optogenetic engineering to probe the molecular choreography of STIM1-mediated cell signaling. *Nat Commun*, 2020. **11**(1): p. 1039.
15. He, L., et al., Near-infrared photoactivatable control of Ca(2+) signaling and optogenetic immunomodulation. *Elife*, 2015. **4**.
16. Kyung, T., et al., Optogenetic control of endogenous Ca(2+) channels in vivo. *Nat Biotechnol*, 2015. **33**(10): pp. 1092–1096.
17. Ishii, T., et al., Light generation of intracellular Ca(2+) signals by a genetically encoded protein BACCS. *Nat Commun*, 2015. **6**: p. 8021.
18. Pham, E., E. Mills, and K. Truong, A synthetic photoactivated protein to generate local or global Ca(2+) signals. *Chem Biol*, 2011. **18**(7): pp. 880–890.
19. He, L., et al., Engineering of a bona fide light-operated calcium channel. *Nat Commun*, 2021. **12**(1): p. 164.
20. Shin, D.M., et al., The TRPCs, Orais and STIMs in ER/PM Junctions. *Adv Exp Med Biol*, 2016. **898**: pp. 47–66.
21. Wang, Y., et al., The calcium store sensor, STIM1, reciprocally controls Orai and CaV1.2 channels. *Science*, 2010. **330**(6000): pp. 105–109.
22. Park, C.Y., A. Shcheglovitov, and R. Dolmetsch, The CRAC channel activator STIM1 binds and inhibits L-type voltage-gated calcium channels. *Science*, 2010. **330**(6000): pp. 101–105.
23. Crosson, S. and K. Moffat, Structure of a flavin-binding plant photoreceptor domain: Insights into light-mediated signal transduction. *Proc Natl Acad Sci U S A*, 2001. **98**(6): pp. 2995–3000.
24. Guntas, G., et al., Engineering an improved light-induced dimer (iLID) for controlling the localization and activity of signaling proteins. *Proc Natl Acad Sci U S A*, 2015. **112**(1): pp. 112–117.

25. Renicke, C., et al., A LOV2 domain-based optogenetic tool to control protein degradation and cellular function. *Chem Biol,* 2013. **20**(4): pp. 619–626.
26. Wu, Y.I., et al., A genetically encoded photoactivatable Rac controls the motility of living cells. *Nature,* 2009. **461**(7260): pp. 104–108.
27. Cosentino, C., et al., Optogenetics. Engineering of a light-gated potassium channel. *Science,* 2015. **348**(6235): pp. 707–710.
28. Alberio, L., et al., A light-gated potassium channel for sustained neuronal inhibition. *Nat Methods,* 2018. **15**(11): pp. 969–976.
29. Dagliyan, O., et al., Engineering extrinsic disorder to control protein activity in living cells. *Science,* 2016. **354**(6318): pp. 1441–1444.
30. Dagliyan, O., N.V. Dokholyan, and K.M. Hahn, Engineering proteins for allosteric control by light or ligands. *Nat Protoc,* 2019. **14**(6): pp. 1863–1883.
31. Dana, H., et al., High-performance calcium sensors for imaging activity in neuronal populations and microcompartments. *Nat Methods,* 2019. **16**(7): pp. 649–657.
32. Zhou, Y., et al., The STIM1-binding site nexus remotely controls Orai1 channel gating. *Nat Commun,* 2016. **7**: p. 13725.

3 Optogenetic Activation of TrkB Signaling

Peiyuan Huang, Zhihao Zhao, Lei Lei, and Liting Duan

CONTENTS

3.1 Introduction 37
3.2 Different Strategies for Optical Control of TrkB Signaling 38
3.3 Constructing Plasmids for Optical Activation of TrkB Signaling 39
3.4 Examining Downstream Signaling Activation by Fluorescent Reporter Assays 41
3.4.1 Cell Culture, Transfection, and Imaging Settings 41
3.4.2 Examining the Activation of MAPK/ERK Signaling 42
3.4.3 Examining the Activation of PI3K/Akt Signaling 42
3.4.4 Examining the Activation of PLCγ1/Ca^{2+} Signaling 45
3.5 PC12 Neurite Growth Assay 46
3.5.1 Cell Culture and Transfection 46
3.5.2 LED Device and Light Stimulation 46
3.5.3 Data Acquisition and Analysis 47
3.5.4 Results 48
3.6 Western Blot 50
3.7 Conclusion 51
References 52

3.1 INTRODUCTION

Brain-derived neurotrophic factor (BDNF)/Tropomyosin receptor kinase B (TrkB) signaling is pivotal for the survival, proliferation, and differentiation of neuronal cells. TrkB is a member of the Trk receptors family that mediates signal transduction in response to growth factors.[1,2] Upon BDNF binding, the transmembrane TrkB receptors undergo homo-interactions, and the cytoplasmic tyrosine kinase domain is trans-autophosphorylated to trigger the downstream signaling pathways.[1] Defective BDNF/TrkB signaling is associated with various diseases, including neuroblastoma,[3] glaucoma,[4] Alzheimer's disease,[5] and Parkinson's disease.[6] Numerous studies have demonstrated that direct administration of BDNF[7] or gene therapy to express BDNF[8] are capable of rescuing the downregulation or dysfunction of TrkB signaling pathways in disease states. However, BDNF ligands also activate another receptor, p75 neurotrophin receptor ($p75^{NTR}$)[9], which has been implicated in cell

DOI: 10.1201/b22823-3

apoptosis.[10] Moreover, the neurotrophic peptides often suffer from short half-lives *in vivo*.[11] Therefore, novel methods for precise modulation of TrkB signaling with high specificity and controllability in time and space are greatly desired to investigate the spatiotemporally dynamic nature of TrkB signaling as well as its roles in health and disease.

Optogenetics is a rapidly evolving and growing biotechnique that provides precise control over various cellular activities.[12–19] By harnessing light-gated ion channels and light-gated protein-protein interactions, optogenetic methods achieve non-invasive, reversible, and specific regulation of targeted protein functions at high spatial and temporal resolution. The expanding pool of optogenetic toolkits offers a variety of choices in applying different photosensitive modules to manipulate diverse cellular processes and has led to many important discoveries. Based on different types of light-gated protein-protein interactions (PPIs), various optogenetic strategies have been exploited to optically control various intracellular signaling pathways.[20–22] Such light-gated PPIs leverage the photosensitive proteins from plants and microbes that undergo intermolecular interactions or intramolecular conformational change when activated by light. Upon light stimulation, photosensitive proteins can homo-associate or hetero-interact with their binding partners. Therefore, the protein of interest (POI) fused with the photosensitive proteins can be directed towards different subcellular locations or undergo homo-interaction or binding with other proteins via specific light-inducible PPIs.

This chapter outlines optogenetic methods based on photosensitive domains, *Arabidopsis thaliana* cryptochrome 2 (CRY2) or *Vaucheria frigida* light-oxygen-voltage domain (AuLOV), to activate intracellular TrkB signaling pathways. The experimental procedures to examine light-induced TrkB activation, including the fluorescent reporter assay and western blot to examine the activation status of downstream signalings and the PC12 neurite growth assay, will be described and summarized.

3.2 DIFFERENT STRATEGIES FOR OPTICAL CONTROL OF TRKB SIGNALING

Endogenous TrkB receptors on the plasma membrane dimerize upon binding to BDNF ligands, transmitting extracellular stimuli into cellular responses by activating downstream signaling pathways. As the homo-interaction of TrkB receptors is central for the signaling activation, photosensitive modules capable of light-mediated homo-interactions can be exploited to optically activate TrkB signaling.

A variety of photosensitive proteins originating from plants and microbes have been engineered and optimized to function in mammalian cells. Among them, cryptochrome 2 from *Arabidopsis thaliana* is extensively studied and widely used in many optogenetic applications. The conserved N-terminal photolyase homology region (PHR) of cryptochrome 2 (CRY2PHR, denoted as CRY2 thereafter) can homo-oligomerize upon blue light stimulation. CRY2, in its photoexcited state, also binds to a basic helix-loop-helix protein *Arabidopsis* CIB1. Such dual characteristics of CRY2[23] have empowered optogenetic control over diverse intracellular activities, such as organelle dynamics,[24,25] and intracellular signaling pathways.[26,27] The light-oxygen-voltage domain of aureochrome 1 from *Vaucheria frigida* (VfAuLOV, denoted as AuLOV thereafter) is another photosensory protein that binds to each other upon blue light stimulation and has been successfully used in controlling receptor dimerization.[28,29] In this chapter, we describe optogenetic systems based on CRY2 and AuLOV photosensitive domains to manipulate TrkB signaling.

TrkB receptor is composed of a ligand-binding extracellular domain, a transmembrane domain, and an intracellular domain that functions as a scaffold to other cytoplasmic kinases upon phosphorylation. Here, only the intracellular domain of TrkB (a.a. 455–822, iTrkB) is used, allowing the optogenetic systems to remain inert toward extracellular ligands. Interactions of signaling proteins may occur at specific subcellular locations to activate the signaling cascades. We applied three types of optogenetic strategies to optically activate TrkB receptors, named OptoTrkB, that employed light-inducible PPIs either on the plasma membrane or in the cell cytosol: (1) cytosolic homo-interactions, (2) membrane-bound homo-interactions, and (3) membrane-recruited homo-interactions (Figure 3.1). In the first OptoTrkB strategy, iTrkB is fused to the photosensitive protein CRY2 or AuLOV and exhibits cytoplasmic distribution. In the second strategy, iTrkB is anchored on the plasma membrane via a membrane-targeting peptide Lyn from the alkylation Src-family kinase. Blue light-induced homo-associations of CRY2 or AuLOV led to the homo-association and consequently the trans-autophosphorylation of iTrkB domains. The third strategy comprises two components that utilize both the light-gated CRY2-CRY2 homo-interactions and heterotypic interaction between CRY2 and the truncated version of CIB1 (a.a. 1–170, CIBN). CIBN is fused to a CAAX motif that targets the plasma membrane. Upon blue light activation, the CRY2-linked iTrkB domain will be recruited from cytosol to the cell membrane via CRY2/CIBN binding and activated by the homo-interactions of CRY2.

3.3 CONSTRUCTING PLASMIDS FOR OPTICAL ACTIVATION OF TRKB SIGNALING

To achieve light-responsive control of TrkB signaling, a series of plasmids encoding CRY2- or AuLOV-based OptoTrkB systems can be constructed by simple ligation and In-Fusion methods (Table 3.1). Here is a detailed description of the plasmid construction procedures.

We employed In-Fusion seamless cloning method (Takara) and DNA ligation method (Thermo Fisher) to construct our OptoTrkB plasmids. The In-Fusion cloning method allows the direct insertion of one or more PCR fragments into linearized vectors. T4 DNA ligase is widely used in linking linearized vectors with DNA fragments digested by restriction enzymes. Here, all the listed plasmid linearization reactions are completed by using Invitrogen™ Anza™ series restriction enzymes.

For plasmid CRY2-mCherry (mCh)-iTrkB construction, the backbone CRY2-mCh-iTrkA[31] was linearized by Bsp1407I and MunI, and the gene of the intracellular domain of human TrkB receptor (iTrkB, NCBI Gene ID: 4915) was inserted into the backbone. For plasmid Lyn-iTrkB-mCh-CRY2, iTrkB-mCh fragment was inserted into the linearized backbone Lyn-iTrkA-mCh-CRY2[31] cut by SacI and Bsp1407I enzymes. To construct AuLOV-mCh-iTrkB, the AuLOV fragment was first amplified by PCR from Lyn-TrkAICD-AuLOV-GFP, a kind gift from Dr. Kai Zhang (University of Illinois at Urbana-Champaign) and inserted into the pmCherry-N1 vector at cutting sites EcoRI and BamHI. The iTrkB gene was then inserted into the AuLOV-mCh linearized by Bsp1407I and MunI to generate AuLOV-mCh-iTrkB. The Lyn-iTrkB-AuLOV-mCh was made by first inserting iTrkB gene into Lyn-TrkAICD-AuLOV-GFP at restriction

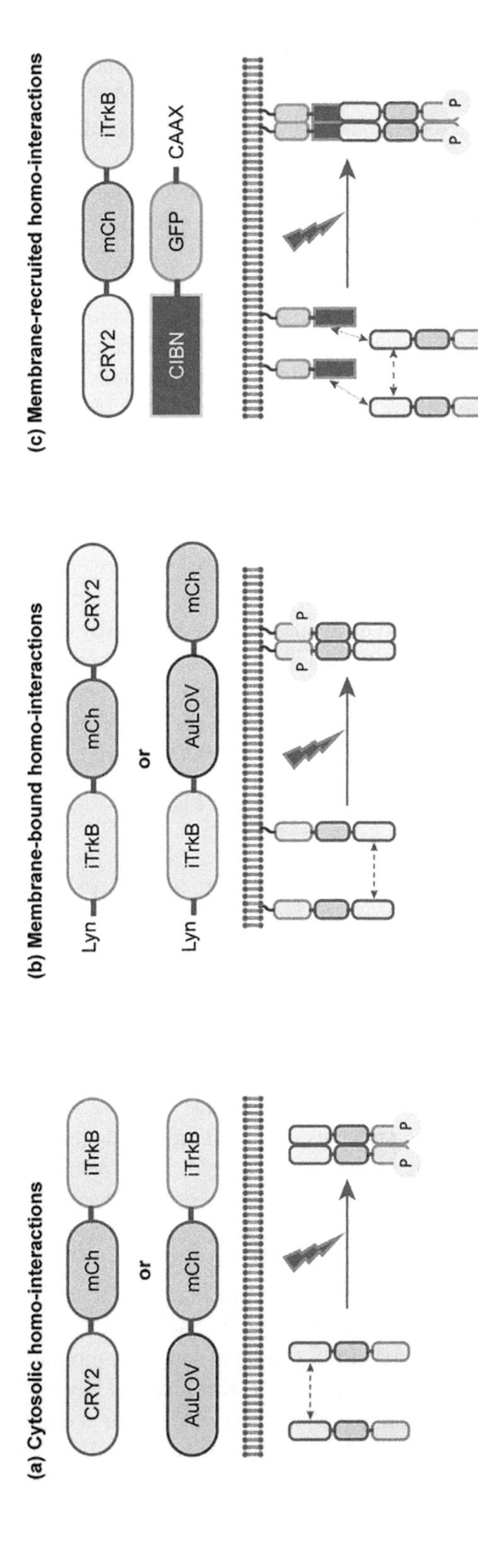

FIGURE 3.1 Different strategies for optical activation of TrkB signaling.[30] (a) CRY2 or AuLOV are fused to the N-terminus of iTrkB. (b) The membrane-targeting sequence Lyn is linked at the N-terminus of iTrkB, with photosensitive proteins CRY2 or AuLOV appended to the C-terminus. (c) CIBN-GFP-CAAX is localized on the plasma membrane, which recruits CRY2-tagged iTrkB upon blue light stimulation.

TABLE 3.1
Primers for Constructing OptoTrkB Systems

Primer No.	Sequence (5′-3′)	To construct
P1	ATGGACGAGCTGTACAAG	CRY2-mCh-iTrkB
P2	AAACGGGCCCTCTAGATTAGCCTAGAATGTCCAGGTAGACCGGAGA	
P3	TCAATGACGATCTCGAGCTCGGCAGTAAGTTGGCAAGACACTCCAAG	Lyn-iTrkB-mCh-CRY2
P4	CTACTGCCCTTGTACAGCTCG	
P5	GGTAGTGCTGGTGATATCAAGTTGGCAAGACACTCCAAG	Lyn-iTrkB-AuLOV-mCh
P6	TCGAGGGCCCGCGGGAGCTCGCCTAGAATGTCCAGGTAGAC	
P7	CTCAAGCTTCGAATTCGCCACCATGCCTGACTACAGTCTCG	AuLOV-mCh
P8	GGCGACCGGTGGATCCTTTCTGCGCAGCATGTTACTGG	
P9	CGCGGGCCCGGGATCCGCCACCATGGTGAGCAAGG	GFP-CAAX
P10	TCTAGAGTCGCGGCCGCCTACATAATTACACACTTTGTC	

enzyme digestion sites of EcoRV and SacI, leading to a GFP-tagged intermediate Lyn-iTrkB-AuLOV-GFP. The final product was obtained by replacing GFP with mCh from pmCherry-N1 via ligation of DNA fragments cut by BamHI and NotI. All primers used in constructing OptoTrkB plasmids are listed in Table 3.1. Other plasmids used in this study can be found in the previous report.[31]

3.4 EXAMINING DOWNSTREAM SIGNALING ACTIVATION BY FLUORESCENT REPORTER ASSAYS

To assess the function of OptoTrkB systems, the live-cell protein translocation imaging assays were employed to monitor the activation of three distinctive hallmark downstream signaling pathways in real time, including the mitogen-activated protein kinase/extracellular signal-regulated kinase (MAPK/ERK) pathway, the phosphatidylinositol 3-kinase/protein kinase B (PI3K/Akt) pathway, and the phospholipase Cγ1 (PLCγ1)/Ca^{2+} signaling.

3.4.1 Cell Culture, Transfection, and Imaging Settings

1. The cell culture for imaging experiments is generally prepared in a confocal dish (SPL #200350, 1.33 cm^2) compatible with microscopic systems. The glass surface of a confocal dish can be coated with 0.01% Poly-L-Lysine (PLL) Solution (Sigma A-005-C) for better cell adhesion.
2. NIH/3T3 cells (ATCC CRL-1658™) were used in this experiment to express the OptoTrkB system and the reporter protein. Nearly 30,000 cells were seeded onto the PLL-coated confocal dish at around 40–50% confluency one day before transfection.

3. Cells were transfected by Lipofectamine™ 3000 Transfection Reagent (Thermo Fisher) following the manufacturer's protocol. Briefly, 0.5 μg total amounts of plasmid DNA was mixed with 1 μL P3000™ enhancer reagent in 50 μL Opti-MEM® I Reduced-Serum Medium (Thermo Fisher). Meanwhile, 1.5 μL Lipofectamine™ 3000 was also prepared in 50 μL Opti-MEM in another Eppendorf 1.5mL microcentrifuge tube. The two solutions were then mixed and incubated at room temperature for 15 min before adding to the cell culture. The amount of DNA, reagents, and Opti-MEM can be adjusted proportionally according to the cell numbers.
4. Cells were allowed to recover for 18–24 hours after transfection and were starved in reduced serum medium (0.5% FBS, DMEM) overnight before imaging experiments.
5. Transfected cells were examined on the 100x objective of a Leica DMi8 S microscope equipped with an on-stage temperature-controlled CO_2 incubation chamber (Tokai Hit GM-8000) and a motorized stage (Prior). Images were captured by a Leica DFC9000 sCMOS Fluorescence Microscope Camera.
6. Time-lapse movies were recorded in two separate channels. 550-nm green light along with a 590/25 nm filter was used for checking the mCh-labeled protein expression of OptoTrkB systems. For examining reporter proteins with green fluorescence, 470-nm blue light was used in combination with a 510/25 nm filter, which could also stimulate blue light-responsive photosensitive proteins. The exposure time was set as 200 ms in both channels.

3.4.2 Examining the Activation of MAPK/ERK Signaling

The nucleus export of kinase translocation reporter (KTR) has been widely employed to report the activation of intracellular signaling pathways.[32] KTR technology measures kinase activities in live cells based on the phosphorylation-regulated nucleo-cytoplasmic translocation. We first used a green fluorescent protein-labeled reporter ERK-KTR-GFP to examine the activation of MAPK/ERK downstream pathways (Figure 3.2a). Upon activation of the Ras/Raf/MEK/ERK signaling cascade, ERK-KTR-GFP will be phosphorylated and be exported outside the nucleus. NIH/3T3 cells were transfected with each of the OptoTrkB systems together with ERK-KTR-GFP and were stimulated by blue light pulses (470 nm, 200-ms pulse at 9.7 W/cm^2) at 5-sec intervals. Upon the delivery of blue light, obvious translocation of ERK-KTR-GFP from the cell nucleus to the cytoplasm was observed in all the 5 proposed systems, indicating the activation of MAPK/ERK signaling (Figure 3.2b). A normalized intensity plot along a dashed line was measured in each image before and after blue light delivery, confirming the significant translocation of ERK-KTR and the signaling activation.

3.4.3 Examining the Activation of PI3K/Akt Signaling

In response to upstream receptor activation, the pleckstrin homology domain of Akt (AktPH) binds to the phospholipid PIP_3 generated by phosphoinositide-3-kinase (PI3K) at the plasma membrane, resulting in further activation of Akt by a

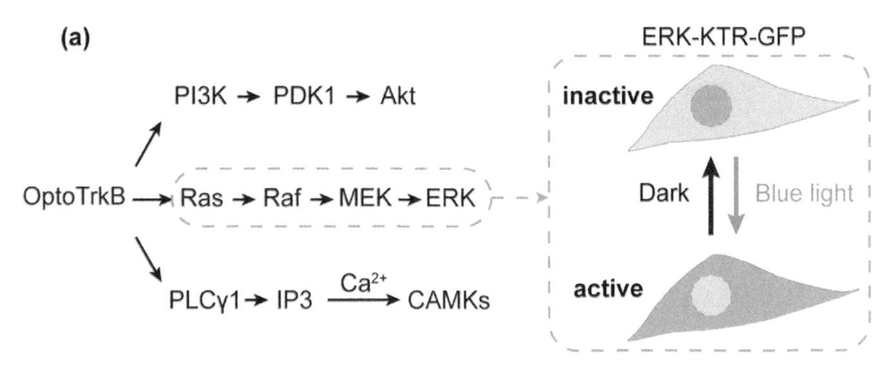

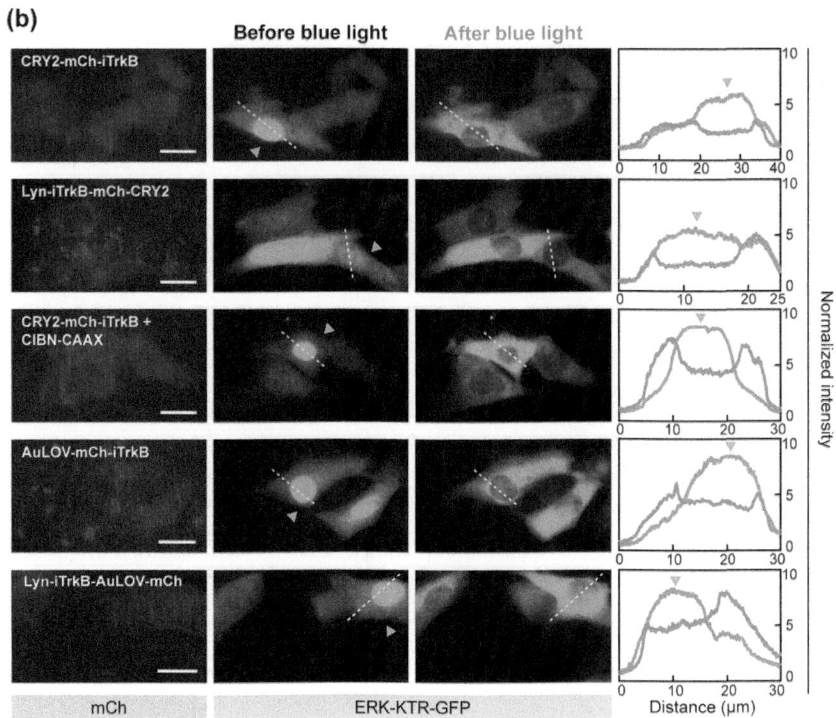

FIGURE 3.2 ERK-KTR translocation assay to examine the activation of MAPK/ERK pathway. (a) Activation of the MAPK/ERK downstream pathway leads to the nuclear export of ERK-KTR-GFP. (b) OptoTrkB systems can activate the downstream MAPK/ERK pathway, as indicated by ERK-KTR-GFP translocation from the nucleus to the cytosol. Right panels show the normalized intensity change along dashed lines before (black line plot) and after blue light stimulation (blue line plot). Scale bars, 20 μm.

phosphoinositide-dependent kinase (PDK).[33] Here, the translocation of the GFP-labeled AktPH domain was used to probe the activities of PI3K/Akt downstream pathway upon OptoTrkB activation (Figure 3.3a). NIH/3T3 cells were transfected with each of the OptoTrkB systems together with AktPH-GFP and were stimulated

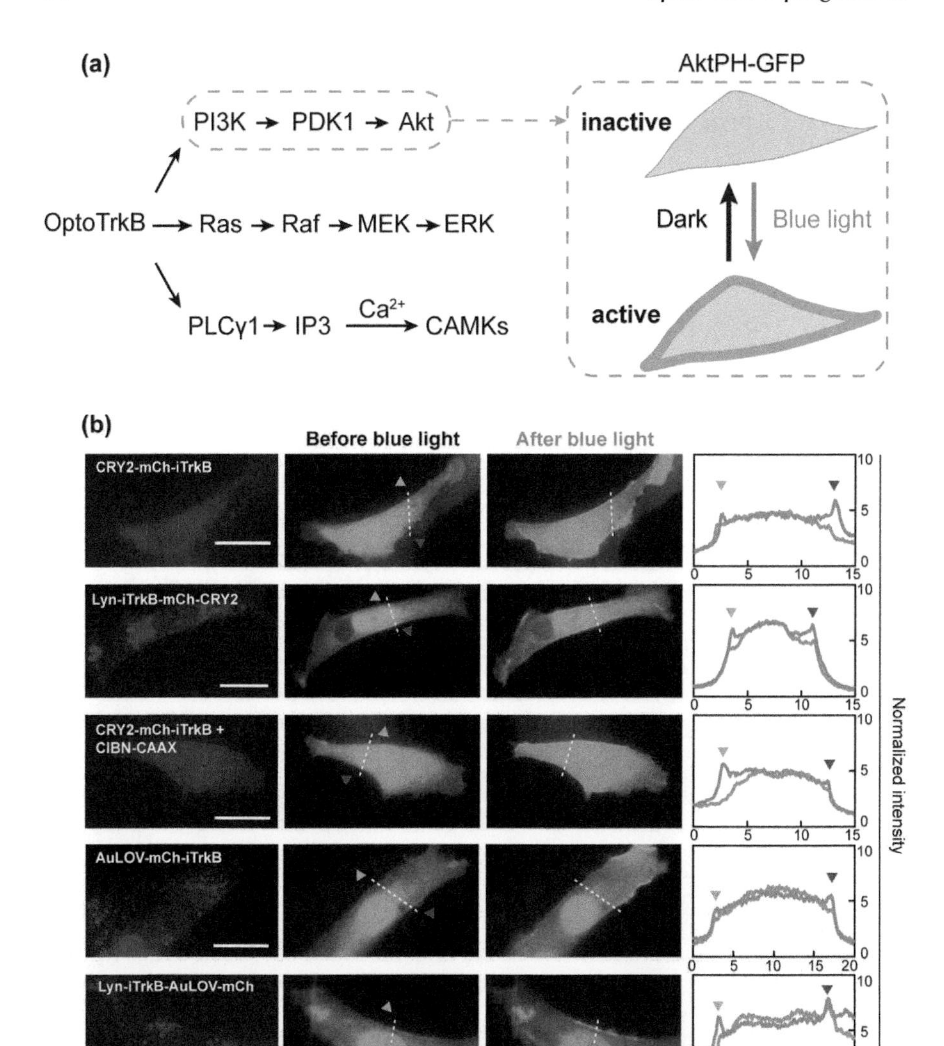

FIGURE 3.3 AktPH-GFP translocation assay to examine the activation of PI3K/Akt pathway. (a) Activation of the PI3K/Akt downstream pathway is indicated by the translocation of AktPH-GFP from cytosol to the plasma membrane. (b) OptoTrkB systems activate the PI3K/Akt pathway, assayed by membrane translocation of AktPH-GFP. Right panels show the normalized intensity change along dashed lines before (black line plot) and after blue light stimulation (blue line plot). Scale bars, 20 μm.

by blue light pulses (470 nm, 200-ms pulse at 9.7 W/cm^2) at 5-sec intervals. In each of the 5 strategies, AktPH-GFP was successfully recruited onto the plasma membrane under blue light illumination with an increased GFP signal at the plasma membrane (Figure 3.3b). The normalized intensity changes can be plotted along dashed lines, where GFP intensity peaks at the membrane positions as indicated by arrows.

3.4.4 Examining the Activation of PLCΓ1/Ca^{2+} Signaling

TrkB activation triggers the PLCγ1/Ca^{2+} signaling, leading to a subsequent increase in intracellular calcium (Ca^{2+}) concentration.[34] The genetically encoded calcium reporter GCaMP6 was employed to probe the upregulation of PLCγ1/Ca^{2+} downstream signaling in response to OptoTrkB activation, where its green fluorescence increases upon the elevation of intracellular calcium concentration.[35] In NIH/3T3 cells transfected with each set of the CRY2-based OptoTrkB strategies with GCaMP6, light stimulation triggered a rapid and significant increase in the green fluorescence signals. Taken together, all the three downstream pathways of TrkB can be activated by OptoTrkB systems, as confirmed by the live-cell imaging assays (Figure 3.4).

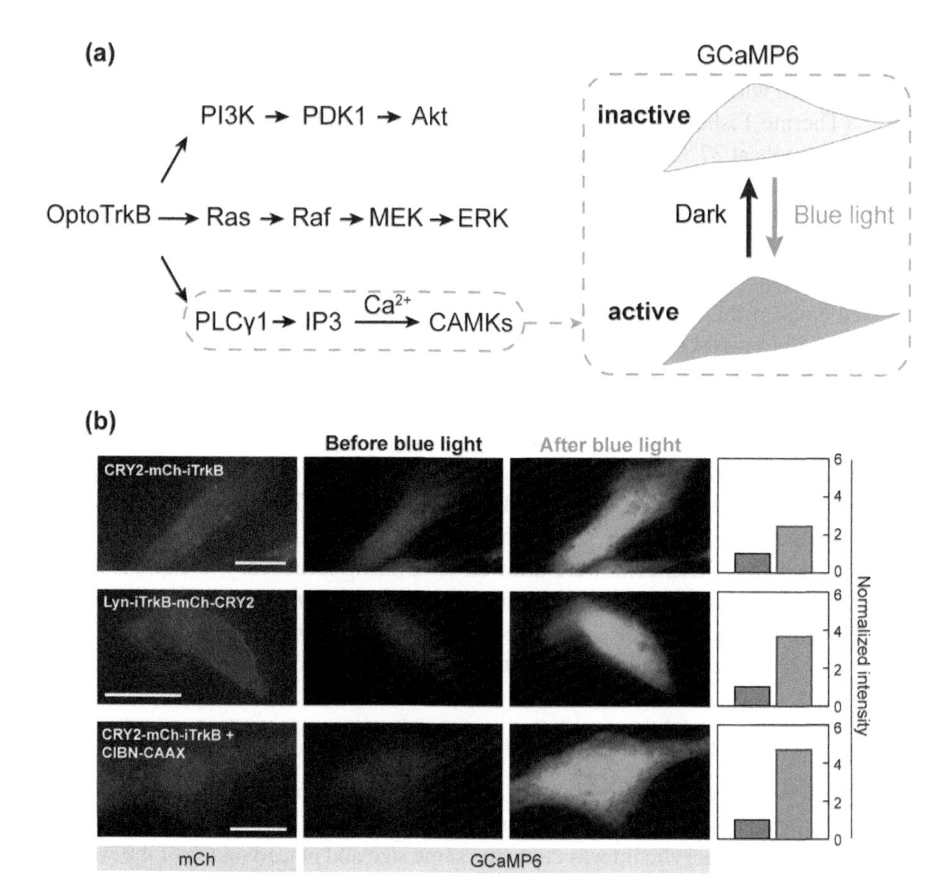

FIGURE 3.4 GCaMP6 assay to examine the activation of phospholipase Cγ1 (PLCγ1)/Ca^{2+} pathway. (a) When OptoTrkB activates downstream PLCγ1/Ca^{2+} signaling pathways, GCaMP6 will emit strong green fluorescence upon the increase of intracellular Ca^{2+} concentration. (b) CRY2-based OptoTrkB systems activate the PLCγ1/Ca^{2+} pathway. Blue light pulses (200 ms, 9.7 W/cm^2) were delivered at 5 s intervals for 5 min. The rise of GCaMP6 signal indicates the activation of Ca^{2+} signaling. Bar graphs in the right panel show the quantified fluorescence changes of each representative cell before and after blue light stimulation. Scale bars, 20 μm.

3.5 PC12 NEURITE GROWTH ASSAY

The rat pheochromocytoma PC12 cell line is widely used as a model for neural differentiation.[36] When exposed to neurotrophin treatment, PC12 cells cease division, extend long branching neurites, and acquire sympathetic neuron electrical properties. Transiently or stably expressed TrkB receptors can mediate functional response to BDNF and stimulate neurite outgrowth in PC12 cells.[37,38] Therefore, the PC12 neurite growth assay can be employed to evaluate the activation of OptoTrkB signaling. The differential efficacies are determined by image-based quantitative analysis of the extent of PC12 neurite growth.

3.5.1 Cell Culture and Transfection

1. PC12 cells (ATCC CRL-1721) were cultured in F-12 K medium supplemented with 15% heat-inactivated horse serum and 2.5% fetal bovine serum (Thermo Fisher). All cell cultures were maintained according to standard protocols at 37°C in a 95% humidified incubator with 5% CO_2.
2. For transient transfection, PC12 cells were seeded 50,000 cells per well in 24-well plates one day before transfection.
3. All cells were transfected with desired DNA plasmids using Lipofectamine™ 3000 reagent according to the manufacturer's protocol. Briefly, 0.25 µg of the plasmid DNA from single-component OptoTrkB systems (CRY2-mCh-iTrkB, Lyn-iTrkB-mCh-CRY2, AuLOV-mCh-iTrkB or Lyn-iTrkB-AuLOV-mCh) was used together with 0.25 µg GFP-CAAX in order to visualize cell morphology with green fluorescence. In the case of CRY2-mCh-iTrkB + CIBN-GFP-CAAX, 0.25 µg of each plasmid was used.
4. For the control groups expressing full length TrkB receptors to be treated with BDNF ligands, 0.25 µg TrkB-mCh and 0.25 µg GFP-CAAX were used together for transfection into each well of PC12 cells.
5. Transfected PC12 cells were allowed for overnight recovery to express the protein, followed by 6 h starvation in pre-neuronal environment of reduced serum medium (F-12K, supplemented with 1.5% horse serum and 0.25% fetal bovine serum) in the dark prior to any stimulation.

3.5.2 LED Device and Light Stimulation

1. To construct a custom-built blue light LED array, a 96 lamp-bead blue LED was placed on the bottom of a plastic rack.
2. A semi-clear acrylic lid was cut to the same size and placed on top of the rack in order to achieve homogeneous light intensity in the illuminated area.
3. The LED circuit is connected to a switched-mode power source that also controls various light on/off illumination modes, e.g., a 5-sec on, 5-sec off light pattern.
4. Light intensity was measured by the Thorlabs power meter.
5. The 24-well cell culture plate was placed onto the LED device. Cells were illuminated with blue light (470nm) at 200 µW/cm^2.
6. An aluminum box was positioned above the LED device to avoid other light interference.

3.5.3 Data Acquisition and Analysis

Measurement of the number of neurites provides a quantitative assessment of PC12 neuronal differentiation and growth.[39] To determine which system can most effectively promote PC12 differentiation, we measured two parameters from the acquired data: the percentage of differentiated PC12 cells among all transfected cells, and the average number of neurite branches in differentiated cells. PC12 cells were counted as differentiated if exhibiting at least one neurite with a length equal to or greater than the cell body diameter. The number of differentiated cells was determined by visual examination of the field and manual counting assisted by the cell counter plug-in of ImageJ software.

1. After 24 h blue light stimulation at 200 μW/cm^2 (5 s on, 5 s off), PC12 cells were first rinsed with cold PBS and then fixed by 4% paraformaldehyde (PFA, diluted from Pierce™ 16% formaldehyde, methanol-free) for 15 min at 37°C.
2. Cell samples were kept immersed in the PBS buffer after the removal of PFA.
3. Fluorescence pictures were obtained from a Leica DMi8 S microscope using either 20× or 40× objective. The membrane-tagged GFP-CAAX or CIBN-GFP-CAAX can be imaged through 470/510 nm ex/em dichroic filters with up to 500 ms exposure time. The mCh-labeled iTrkB protein distribution can be visualized through 550/590 nm filters.
4. For cell counting, firstly, transfected cells were identified in the GFP channel images by the cell counter of ImageJ to record the total number of transfected cells.
5. All outgrowing neurites longer than the cell body diameters in the field of image were counted by the cell counter.
6. As shown in Table 3.2, the total number of counted cells and differentiated cells with neurite growth was recorded for each group in three independent replicates. Cells bearing at least one neurite with a length equal to or greater than the cell body diameter were counted as cells with neurite growth (Neurite +) among total transfected cells (Total #).

TABLE 3.2
Counting Numbers for the Quantification of PC12 Cell Differentiation

Group		Set 1		Set2		Set3		Average
		Total #	Neurite+	Total #	Neurite+	Total #	Neurite+	percentage
CRY2-mCh-iTrkB	LIGHT	331	77	189	30	316	53	19.14%
	DARK	403	22	375	17	374	21	5.21%
Lyn-iTrkB-mCh-CRY2	LIGHT	342	69	336	56	308	61	18.86%
	DARK	356	26	315	19	350	25	6.86%
CRY2-mCh-iTrkB + CIBN-GFP-CAAX	LIGHT	295	162	285	105	271	109	44.18%
	DARK	245	18	310	20	294	20	6.83%

(Continued)

TABLE 3.2
(Continued)

Group		Set 1		Set2		Set3		Average
		Total #	Neurite+	Total #	Neurite+	Total #	Neurite+	percentage
AuLOV-mCh-iTrkB	LIGHT	280	33	275	30	344	41	11.57%
	DARK	289	16	350	20	358	14	5.02%
Lyn-iTrkB-AuLOV mCh	LIGHT	226	41	261	37	330	49	15.54%
	DARK	265	17	345	19	372	25	6.21%
CIBN-GFP-CAAX	LIGHT	153	9	163	6	265	11	4.48%
	DARK	180	9	334	17	235	11	4.94%
BDNF treatment	0 ng/mL	345	22	312	32	360	23	7.57%
	25 ng/mL	324	119	368	141	364	144	38.26%
	50 ng/mL	338	109	331	123	339	118	34.72%
	75 ng/mL	359	158	317	137	323	117	41.24%
	100 ng/mL	341	162	327	111	377	144	39.90%

3.5.4 Results

Varied extents of neurite outgrowth were observed among all light-activated OptoTrkB groups (Figure 3.5a). By contrast, cells expressing OptoTrkB constructs but kept in the dark did not have noticeable neurites, proving that the differentiation of PC12 cells was induced by light-induced activation of TrkB signaling. As control, cells expressing only CIBN-GFP-CAAX had no obvious neurite outgrowth either under light illumination or in the dark.

In the presence of light stimulation, cells expressing either CRY2- or AuLOV-based OptoTrkB systems had more neurite outgrowth than the CIBN-GFP-CAAX control group. Particularly, in the group of CRY2-mCh-iTrkB + CIBN-GFP-CAAX, 44% of cells grew neurites upon blue light stimulation, which is the highest percentage of differentiated cells among all groups. Meanwhile, similar portions of cell differentiation under light stimulation were obtained in the groups of CRY2-mCh-iTrkB (19%), Lyn-iTrkB-mCh-CRY2 (19%), or Lyn-AuLOV-mCh-iTrkB (16%). Twelve percent of cells transfected with AuLOV-mCh-iTrkB had neurite outgrowth. It is slightly larger than the light-illuminated control group CIBN-GFP-CAAX, which only had 5% of cells showing noticeable neurite growth.

We also found out that the efficacies of our optogenetic systems are comparable to the BDNF-mediated TrkB signaling. In the transfected PC12 cells expressing full-length TrkB receptors, neurite outgrowth was induced by various concentrations of BDNF treatment. CRY2-mCh-iTrkB + CIBN-GFP-CAAX, as the strategy possessing best efficacy in promoting PC12 differentiation among all OptoTrkB groups, had a similar level of PC12 cell neurite outgrowth to that induced by 75 ng/mL BDNF treatment. In addition, no difference in the average numbers of neurite branches were observed among all groups of differentiated cells (Figure 3.5c).

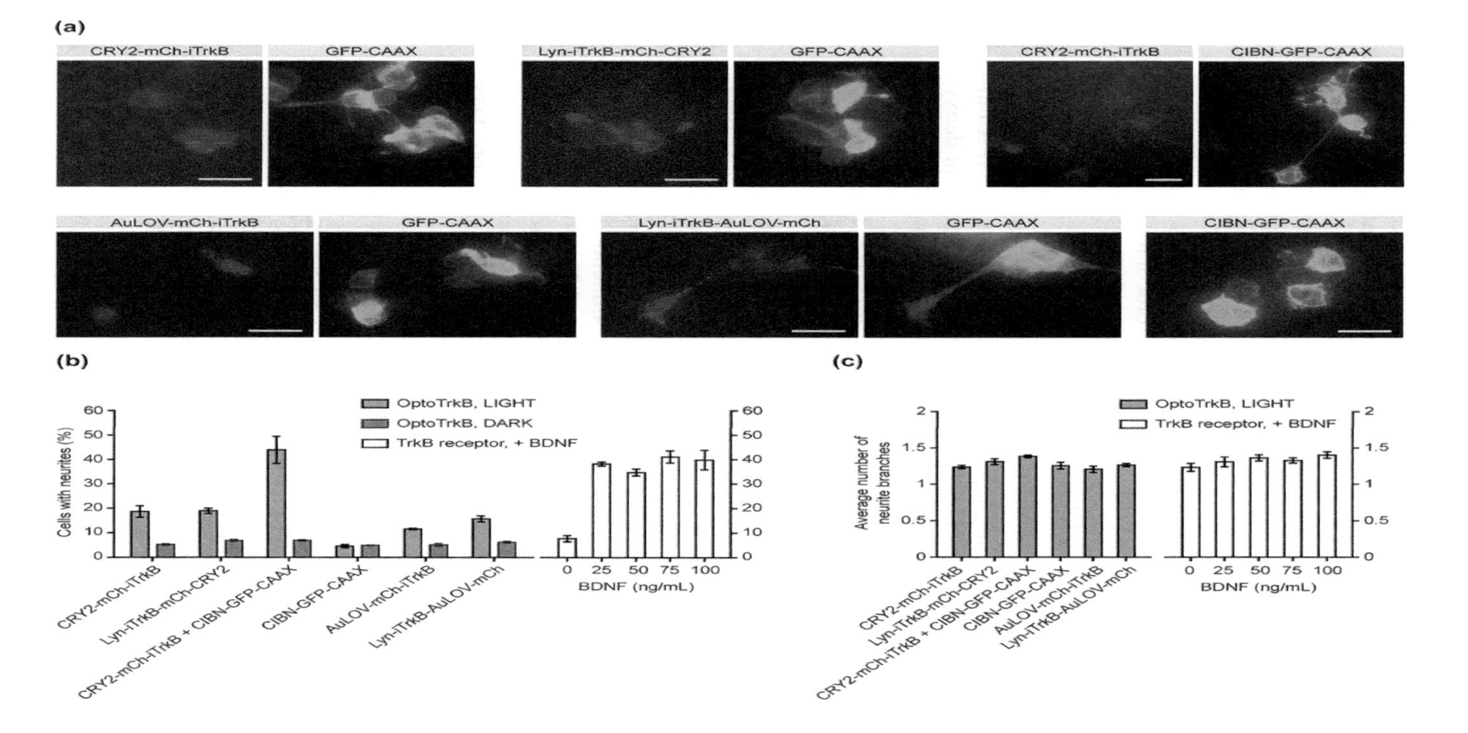

FIGURE 3.5 CRY2-mCh-iTrkB + CIBN-GFP-CAAX strategy shows the best efficiency in promoting PC12 neurite growth. (a) Blue light-activated OptoTrkB systems promote neurite growth in PC12 cells in the absence of BDNF. In PC12 cells expressing each of the OptoTrkB systems, noticeable neurite growth was observed. In contrast, PC12 cells only transfected with CIBN-GFP-CAAX did not show obvious neurite growth. Scale bars, 20 μm. (b) Quantification of the percentage of cells bearing neurites. Cells were counted as differentiated cells if exhibiting at least one neurite longer than cell body diameter in length. CRY2-mCh-iTrkB + CIBN-GFP-CAAX is the most effective strategy to promote neurite outgrowth, similar to that induced by 75 ng/mL BDNF treatment. (c) Quantification of the average number of neurite branches in differentiated PC12 cells. No difference was observed among all groups of differentiated cells. Results represent mean ± SEM in (b) and (c) of over 300 cells in each OptoTrkB group across three independent sets of experiments.

3.6 WESTERN BLOT

To assess TrkB signaling activation among different OptoTrkB strategies, we employed the western blot to examine the levels of phosphorylated ERK1/2 and phosphorylated Akt. PC12 cells (ATCC CRL-1721) were cultured in F-12 K medium supplemented with 15% heat-inactivated horse serum and 2.5% fetal bovine serum (Thermo Fisher Scientific). Cells were transfected by electroporation using the Amaxa Nucleofector II (Lonza). Briefly, cells were added to a suspension of DNA in electroporation buffer (7 mM ATP, 11.7 mM $MgCl_2$, 86 mM KH_2PO_4, 13.7 mM $NaHCO_3$, 1.9 mM glucose), transferred to a 2-mm electroporation cuvette (Fisher Scientific), and subjected to the manufacturer-provided protocol for PC12 cells. Cells were allowed to recover in culture medium for 40 h, followed by 8-h serum starvation prior to the western blot experiments carried out as follows:

1. Cells were exposed to continuous blue light at 400 μW/cm^2 for 20 min before being moved to ice, rinsed with cold PBS, and lysed in RIPA buffer (25 mM Tris HCl, 150 mM NaCl, 1% Triton X-100, 1% sodium deoxycholate, 0.1% SDS) supplemented with protease and phosphatase inhibitor cocktails (Roche 04906837001 and 04693132001).
2. Cell lysates were then scraped off from the plate by a cell lifter and collected into 1.5 mL microcentrifuge tubes, followed by 60 min incubation on ice for sufficient lysis.
3. After centrifugation at 12,000 x g at 4°C for 15 min, clear supernatants were transferred to pre-cooled microcentrifuge tubes. Cell pellets were discarded.
4. Protein concentration of samples were determined using the BCA Protein Assay Kit (Thermo Fisher Scientific) according to the manual.
5. Protein samples were mixed in a 3:1 ratio with 4x Laemmli buffer (Bio-Rad #1610747) containing 2-mercaptoethanol. It is recommended to normalize protein samples with RIPA buffer before addition of the laemmli sample buffer so that equal volume for all samples can be loaded into gels.
6. Protein samples were boiled at 95°C for 5 min for denaturation.
7. Fifteen μg of protein samples were subjected to SDS-PAGE using 10% TGX Stain-Free polyacrylamide gels (BioRad) in a standard Tris-glycine buffer. Voltage was set at 80V for the first 20 min to get samples lined up in the gel and then increased to 120V for about 90 min to separate protein completely.
8. After separation, protein was transferred to a PVDF membrane using a Trans-Blot Turbo semi-dry transfer apparatus (Bio-rad) at 2.5A constant current, up to 25V for 8 min. PVDF membranes were rinsed briefly with TBS-T buffer, followed by blocking in 5% non-fat milk in TBS-T for 1 h, shaking at room temperature. The PVDF membrane should be kept wet and clean all the time.
9. Primary antibodies were all obtained from Cell Signaling Technology (CST) and diluted in TBS-T buffer: 1:500 for anti-pAKT (CST 9275), 1:1000 for anti-AKT (CST 9272), anti-pERK1/2 (CST 9101), anti-ERK1/2 (CST 9102), and anti-GAPDH (CST 2118).
10. After overnight incubation at 4°C, the blot was subject to 5 min washing in the TBS-T buffer three times. The blot was then incubated at room

temperature with HRP-conjugated secondary antibody (CST 7074) diluted in TBS-T by 1:2000. After three times' washing in TBS-T, protein bands were visualized by Clarity™ Western ECL Substrate (BioRad 1705061) using a ChemiDoc imaging system.

11. Western blots were quantified by ImageJ by densitometric analysis. Briefly, all bands (pERK, total ERK, pAkt, total Akt) were first normalized to the internal loading control GAPDH bands. The relative phosphorylation levels were then quantified by calculating the phosphorylated bands over corresponding total protein levels.

The results of western blot revealed that all groups had minimal levels of pERK and pAkt in the absence of blue light illumination, although Lyn-iTrkB-mCh-CRY2 and Lyn-iTrkB-AuLOV-mCh groups gained slightly higher dark activation. In contrast, blue light stimulation increased pERK and pAkt levels in all OptoTrkB groups. There was an even greater rise in the CRY2-mCh-iTrkB and CRY2-mCh-iTrkB + CIBN-GFP-CAAX groups. As a negative control, cells transfected with only CIBN-GFP-CAAX did not display pERK or pAkt increase when illuminated. Quantification of western blots confirmed that CRY2-mCh-iTrkB + CIBN-GFP-CAAX strategy leads to the strongest activation of TrkB downstream MAPK/ERK and PI3K/Akt signaling pathways, as shown in Figure 3.6b and 3.6c.

3.7 CONCLUSION

TrkB signaling is critical for neuronal cell survival, proliferation, and differentiation. Conventionally, the control of intracellular signaling pathways relies on the treatment of growth factors or inhibitors. The optogenetic designs utilize the intracellular domain of TrkB receptors and bypass side effects from agonists and antagonists that unspecifically act on other extracellular receptor domains. Moreover, such optical regulation provides new solutions to modulate TrkB signaling with spatial and temporal precision. Strategies based on either CRY2/CIBN or AuLOV photosensory domains are able to trigger TrkB signaling activation, indicating that the design principles of OptoTrkB can be further combined with other optogenetic systems for different experimental settings and application purposes.

The effective, non-invasive optical activation of TrkB signaling is validated by live-cell imaging, western blot, and PC12 neurite growth assays. The three downstream pathways, including MAPK/ERK, PI3K/Akt, and PLCγ1/Ca^{2+} pathways, are efficiently upregulated upon light-triggered activation of OptoTrkB systems. Among the listed strategies, membrane-recruitment of the intracellular domain of TrkB receptor via CRY2/CIBN hetero-interactions (CRY2-mCh-iTrkB + CIBN-CAAX) proves to be the most efficient to activate TrkB signaling and promote PC12 neurite outgrowth. The efficiency of optical activation was comparable to 75 ng/mL BDNF treatment in the PC12 neurite growth assay. OptoTrkB systems can be further optimized based on engineered photosensitive proteins, such as CRY2 mutants with an increased tendency of oligomerization. We envision that our strategy holds great promise for investigating the roles of dynamic TrkB signaling events in various physiological and pathological models.

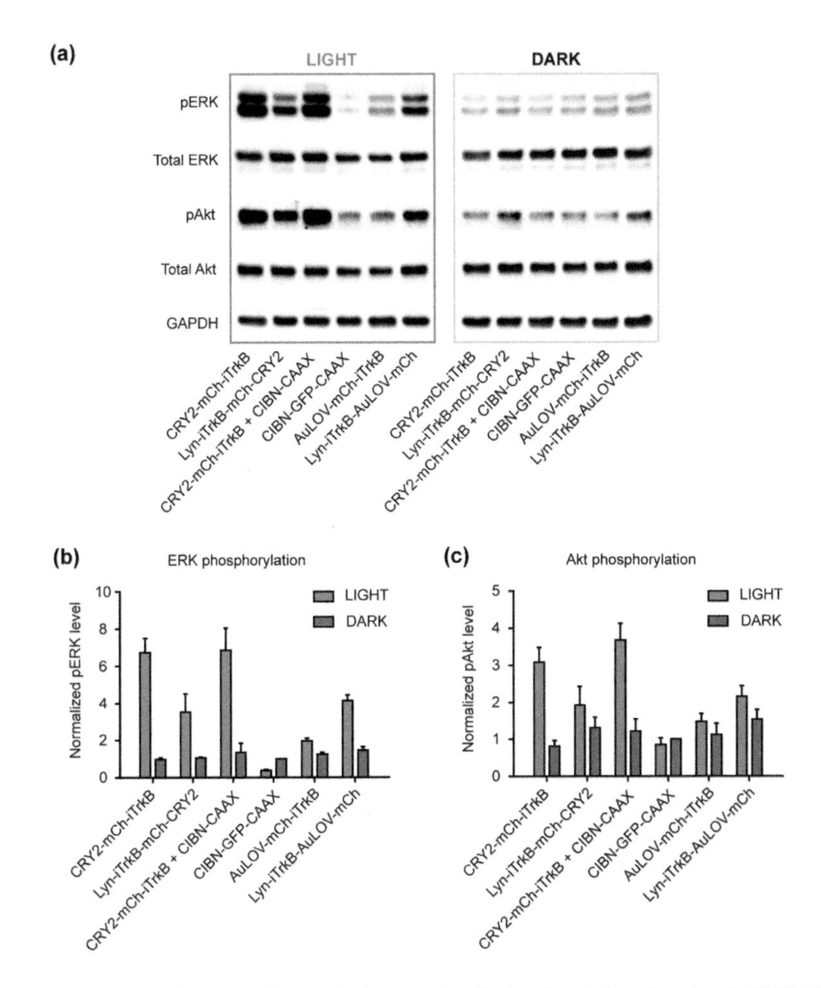

FIGURE 3.6 (a) Western blot analysis to probe the levels of phosphorylated ERK (pERK) and phosphorylated Akt (pAkt) showed different activation strength of MAPK/ERK and PI3K/Akt downstream pathways. The group of CRY2-mCh-iTrkB + CIBN-CAAX can induce the highest levels of pERK and pAkt. (b-c) Quantification of ERK and Akt phosphorylation level by densitometry analysis. pERK and pAkt levels were normalized to total ERK or total Akt levels. Results were further normalized to the calculated pERK or pAkt level in the CIBN-GFP-CAAX control group kept in the dark. Plotted values represent the mean ± SEM of three technical replicates across two independent biological experiments.

REFERENCES

1. Huang, E. J. & Reichardt, L. F. Trk receptors: roles in neuronal signal transduction. *Annu. Rev. Biochem.* **72**, 609–642 (2003).
2. Patapoutian, A. & Reichardt, L. F. Trk receptors: mediators of neurotrophin action. *Curr. Opin. Neurobiol.* **11**, 272–280 (2001).
3. Nakagawara, A., Azar, C. G., Scavarda, N. J. & Brodeur, G. M. Expression and function of TRK-B and BDNF in human neuroblastomas. *Mol. Cell. Biol.* **14**, 759–767 (1994).

4. Mysona, B. A., Zhao, J. & Bollinger, K. E. Role of BDNF/TrkB pathway in the visual system: therapeutic implications for glaucoma. *Expert Rev. Ophthalmol.* **12**, 69–81 (2017).
5. Ferrer, I. et al. BDNF and full-length and truncated TrkB expression in Alzheimer disease. Implications in therapeutic strategies. *J. Neuropathol. Exp. Neurol.* **58**, 729–739 (1999).
6. Fenner, M. E., Achim, C. L. & Fenner, B. M. Expression of full-length and truncated trkB in human striatum and substantia nigra neurons: implications for Parkinson's disease. *J. Mol. Histol.* **45**, 349–361 (2014).
7. Domenici, L. et al. Rescue of Retinal Function by BDNF in a Mouse Model of Glaucoma. *PLoS ONE* **9**, e115579 (2014).
8. Martin, K. R. G. et al. Gene therapy with brain-derived neurotrophic factor as a protection: retinal ganglion cells in a Rat Glaucoma model. *Invest. Ophthalmol. Vis. Sci.* **44**, 4357 (2003).
9. Kaplan, D. R. & Miller, F. D. Neurotrophin signal transduction in the nervous system. *Curr. Opin. Neurobiol.* **10**, 381–391 (2000).
10. Lee, K.-F. et al. Targeted mutation of the gene encoding the low affinity NGF receptor p75 leads to deficits in the peripheral sensory nervous system. *Cell* **69**, 737–749 (1992).
11. Aloe, L., Rocco, M. L., Bianchi, P. & Manni, L. Nerve growth factor: from the early discoveries to the potential clinical use. *J. Translat. Med.* **10** (2012).
12. Kolar, K. & Weber, W. Synthetic biological approaches to optogenetically control cell signaling. *Curr. Opin. Biotechnol.* **47**, 112–119 (2017).
13. Johnson, H. E. & Toettcher, J. E. Illuminating developmental biology with cellular optogenetics. *Curr. Opin. Biotechnol.* **52**, 42–48 (2018).
14. Yu, D. et al. Optogenetic activation of intracellular antibodies for direct modulation of endogenous proteins. *Nat. Methods* **16**, 1095–1100 (2019).
15. Zhou, Y. et al. A small and highly sensitive red/far-red optogenetic switch for applications in mammals. *Nat. Biotechnol.* **40**, 262–272 (2022).
16. Liu, R. et al. Optogenetic control of RNA function and metabolism using engineered light-switchable RNA-binding proteins. *Nat. Biotechnol.* **40**, 779–786 (2022).
17. Nguyen, N. T., Ma, G., Zhou, Y. & Jing, J. Optogenetic approaches to control Ca-modulated physiological processes. *Curr. Opin. Physiol.* **17**, 187–196 (2020).
18. Leopold, A. V. & Verkhusha, V. V. Light control of RTK activity: from technology development to translational research. *Chem. Sci.* **11**, 10019–10034 (2020).
19. Passmore, J. B., Nijenhuis, W. & Kapitein, L. C. From observing to controlling: Inducible control of organelle dynamics and interactions. *Curr. Opin. Cell Biol.* **71**, 69–76 (2021).
20. Tischer, D. & Weiner, O. D. Illuminating cell signalling with optogenetic tools. *Nat. Rev. Mol. Cell Biol.* **15**, 551–558 (2014).
21. Zhang, K. & Cui, B. Optogenetic control of intracellular signaling pathways. *Trends Biotechnol.* **33**, 92–100 (2015).
22. Huang, P., Zhao, Z. & Duan, L. Optogenetic activation of intracellular signaling based on light-inducible protein-protein homo-interactions. *Neural Regeneration Res.* **17**, 25–30 (2022).
23. Che, D. L., Duan, L., Zhang, K. & Cui, B. The dual characteristics of light-induced cryptochrome 2, homo-oligomerization and heterodimerization, for optogenetic manipulation in mammalian cells. *ACS Synth. Biol.* **4**, 1124–1135 (2015).
24. Wittmann, T., Dema, A. & van Haren, J. Lights, cytoskeleton, action: Optogenetic control of cell dynamics. *Curr. Opin. Cell Biol.* **66**, 1–10 (2020).
25. Song, Y. et al. Light-inducible deformation of mitochondria in live cells. *Cell Chem. Biol.* **29**, 109–119 (2021).
26. Repina, N. A., Rosenbloom, A., Mukherjee, A., Schaffer, D. V. & Kane, R. S. At light speed: advances in optogenetic systems for regulating cell signaling and behavior. *Annu. Rev. Chem. Biomol. Eng.* **8**, 13–39 (2017).

27. Crossman, S. H. & Janovjak, H. Light-activated receptor tyrosine kinases: designs and applications. *Curr. Opin. Pharmacol.* **63**, 102197 (2022).
28. Grusch, M. et al. Spatio-temporally precise activation of engineered receptor tyrosine kinases by light. *EMBO J.* **33**, 1713–1726 (2014).
29. Khamo, J. S., Krishnamurthy, V. V., Chen, Q., Diao, J. & Zhang, K. Optogenetic delineation of receptor tyrosine kinase subcircuits in PC12 Cell differentiation. *Cell Chem. Biol.* **26**, 400–410.e3 (2019).
30. Huang, P. et al. Optical activation of TrkB signaling. *J. Mol. Biol.* **432**, 3761–3770 (2020).
31. Duan, L. et al. Optical activation of TrkA signaling. *ACS Synth. Biol.* **7**, 1685–1693 (2018).
32. Regot, S., Hughey, J. J., Bajar, B. T., Carrasco, S. & Covert, M. W. High-sensitivity measurements of multiple kinase activities in live single cells. *Cell* **157**, 1724–1734 (2014).
33. Hemmings, B. A. & Restuccia, D. F. PI3K-PKB/Akt pathway. *Cold Spring Harb. Perspect. Biol.* **4**, a011189 (2012).
34. Minichiello, L. et al. Mechanism of TrkB-mediated hippocampal long-term potentiation. *Neuron* **36**, 121–137 (2002).
35. Chen, T.-W. et al. Ultrasensitive fluorescent proteins for imaging neuronal activity. *Nature* **499**, 295–300 (2013).
36. Teng, K., Angelastro, J., Cunningham, M. & Greene, L. Cultured PC12 CellsA model for neuronal function, differentiation, and survival. *Cell Biol.* 171–176 (2006). doi:10.1016/b978-012164730-8/50022-8
37. Squinto, S. P. et al. trkB encodes a functional receptor for brain-derived neurotrophic factor and neurotrophin-3 but not nerve growth factor. *Cell* **65**, 885–893 (1991).
38. Iwasaki, Y., Ishikawa, M., Okada, N. & Koizumi, S. Induction of a distinct morphology and signal transduction in TrkB/PC12 cells by nerve growth factor and brain-derived neurotrophic factor. *J. Neurochem.* **68**, 927–934 (1997).
39. Harrill, J. A. & Mundy, W. R. Quantitative assessment of neurite outgrowth in PC12 cells. *Methods Mol. Biol.* **758**, 331–348 (2011).

4 Spatiotemporal Modulation of Neural Repair

Huaxun Fan, Qin Wang, Kai Zhang, and Yuanquan Song

CONTENTS

4.1 Introduction ... 55
4.2 Design of optoRaf and optoAKT ... 56
4.3 Validation of Optogenetic Systems *in vitro* ... 56
4.4 Activation of the Optogenetic System in Flies ... 60
 4.4.1 Generation of the *UAS-optoRaf/AKT* Transgenic Flies ... 60
 4.4.2 Activation of optoAKT and optoRaf *in vivo* ... 60
 4.4.3 Inactivation of optoAKT and optoRaf ... 62
4.5 Local Optogenetic Stimulation to Guide Regenerating Axons ... 62
 4.5.1 Fly Sensory Neuron Injury Model ... 62
 4.5.2 Whole-Field Light Stimulation Promotes Axon Regeneration but Fails to Improve Pathfinding ... 64
 4.5.3 Local Illumination of an Axon Branch Leads to Preferred Regeneration ... 65
4.6 Concluding Remarks and Perspective ... 67
Acknowledgments ... 68
References ... 68

4.1 INTRODUCTION

Inadequate neuroregeneration remains a major roadblock toward functional recovery after nervous system damage such as stroke, spinal cord injury (SCI), and multiple sclerosis, causing paralysis in an estimated 1.7% of the United States population [1]. The neurotrophic signaling pathway, which regulates neurogenesis during embryonic development, represents an essential intrinsic regenerative machinery [2]. Elimination of the Pten phosphatase, an endogenous brake for neurotrophic signaling, yields axonal regeneration [3]. However, significant pain generation induced by the efficacious dose of neurotrophins has severely limited their potential as regenerative medicine [4].

An important feature of the neurotrophin signaling pathway is that the functional outcome depends on signaling kinetics [5] and subcellular localization [6]. Indeed, neural regeneration from damaged neurons is synergistically regulated by multiple

DOI: 10.1201/b22823-4

signaling circuits in space and time. Pharmacological application of neurotrophins leads to widespread chronic activation of neurotrophic signaling. Moreover, the PLCγ signaling, a subcircuit of the neurotrophic signaling, is associated with pain generation. Therefore, there is an urgent need to develop a tool that enables precise spatiotemporal manipulation of the regenerative-specific subcircuits of the neurotrophic signaling, such as the Raf/MEK/ERK and AKT signaling pathways.

The emerging optogenetic approach uses light-sensitive channelrhodopsins to control the ion flow of specific neurons and has significantly increased the spatiotemporal resolution of neuronal manipulation. Channelrhodopsins and their derivatives, however, do not confer channel-specific information because of their low ion selectivity. Opsin-free optogenetics extends the application of photoactivatable proteins into optical control of specific cell signalings and represents a promising strategy for precision medicine [7]. As demonstrated in this chapter, we use two light-sensitive optogenetic tools, optoRaf and optoAKT, to control the activity of the corresponding Raf/MEK/ERK and AKT signaling with high spatial and temporal precision, obviating the need of neurotrophins while eliciting robust and guided axon regeneration in transgenic flies.

4.2 DESIGN OF OPTORAF AND OPTOAKT

To specifically control ERK and AKT signaling pathways with a high spatiotemporal resolution, optogenetic systems named optoRaf and optoAKT were developed [8, 9]. A photoactivatable protein pair consisting of the N-terminus of cryptochrome-interacting basic-helix-loop-helix (CIBN) and the photolyase homology region of cryptochrome 2 (CRY2PHR, abbreviated as CRY2) that undergoes light-induced heterodimerization, was used to control the activity of target proteins in these systems [10]. CIBN was anchored on the plasma membrane and CRY2 was fused with c-Raf and PH-domain-truncated AKT (AKTΔPH) to activate the ERK and AKT signaling pathways respectively. To construct a single-transcript system, we inserted a p2A self-cleavage peptide between two functional parts [11, 12] (Figure 4.1a). Upon blue light illumination, the fusion protein complexes will be recruited to the plasma membrane (PM), which leads to the activation of downstream ERK and AKT pathways (Figure 4.1b).

4.3 VALIDATION OF OPTOGENETIC SYSTEMS *IN VITRO*

We first determined the membrane translocation mediated by CRY2-CIBN in cultured mammalian cells. HEK293 cells that express optoRaf or optoAKT show green fluorescence (from CIBN2-GFP-CaaX) on the plasma membrane and red fluorescence in the cytoplasm (from CRY2-mCherry-Raf or CRY2-mCherry-AKT) (Figure 4.2a-b). Upon blue light stimulation, cytosolic CRY2-fusion proteins translocate to the plasma membrane. Membrane-bound CRY2 fusion proteins spontaneously re-disperse into the cytoplasm after switching off the light within minutes (Figure 4.2a-b). Multiple cycles of shuttling between the plasma membrane and the cytoplasm can be resolved by live-cell imaging (Figure 4.2c). Fusion of Raf and AKT does not affect light-induced heterodimerization of CRY2 and CIBN. The

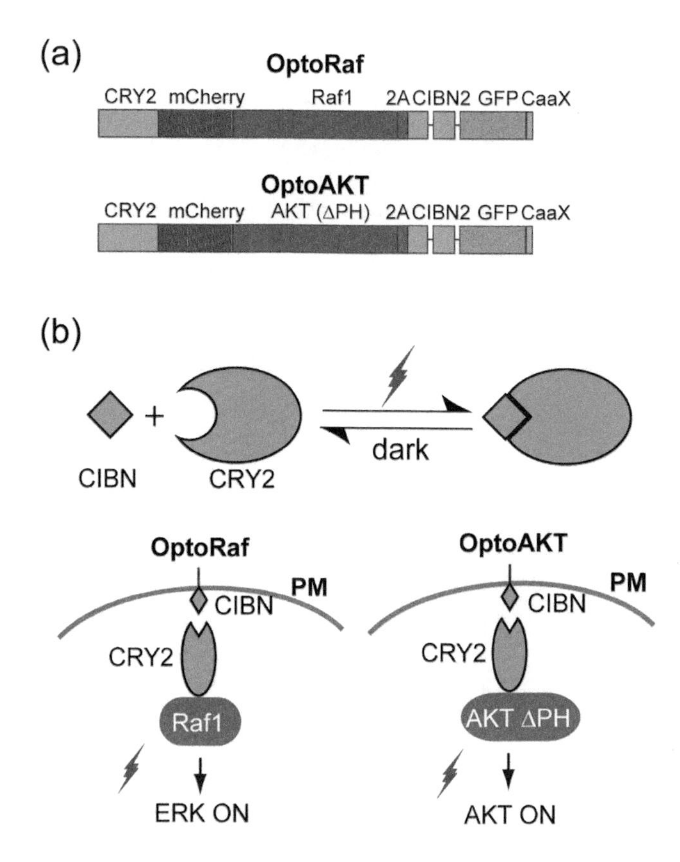

FIGURE 4.1 Scheme for optoRaf and optoAKT. (a) Construct design of optoRaf and optoAKT. A P2A peptide is inserted between CRY2-mCherry-Raf1 (or AKT) and CIBNx2-GFP-CaaX to ensure the bicistronic expression of two fusion proteins. (b) Scheme depicting the working mechanism of optoRaf and AKT. Blue light induces CRY2 and PM-resident CIBN heterodimerization, which recruits Raf (or AKT) to the plasma membrane and activates the downstream signaling pathways. Panel adapted from [12] (CC BY 4.0).

chimeric proteins show similar association and dissociation kinetics of CRY2 and CIBN alone.

To further confirm that light-induced membrane translocation of signaling factors activates their corresponding signaling pathways, we probed the activity of downstream signaling molecules with imaging and biochemical assays. Cells co-transfected with ERK-GFP and optoRaf showed nuclear translocation of ERK-EGFP (downstream readout for Raf activation), whereas cells co-transfected with FOXO3-GFP and optoAKT displayed nuclear export of FOXO3-EGFP (downstream of AKT). These results confirmed that optoRaf and optoAKT activate the Raf/MEK/ERK and AKT signaling pathways, respectively, upon blue light illumination (Figure 4.2d-e). Western blot analysis revealed that pERK and pAKT increased within 10 min illumination and dropped back to the basal level within 30 min after switching off the light (Figure 4.3a-d).

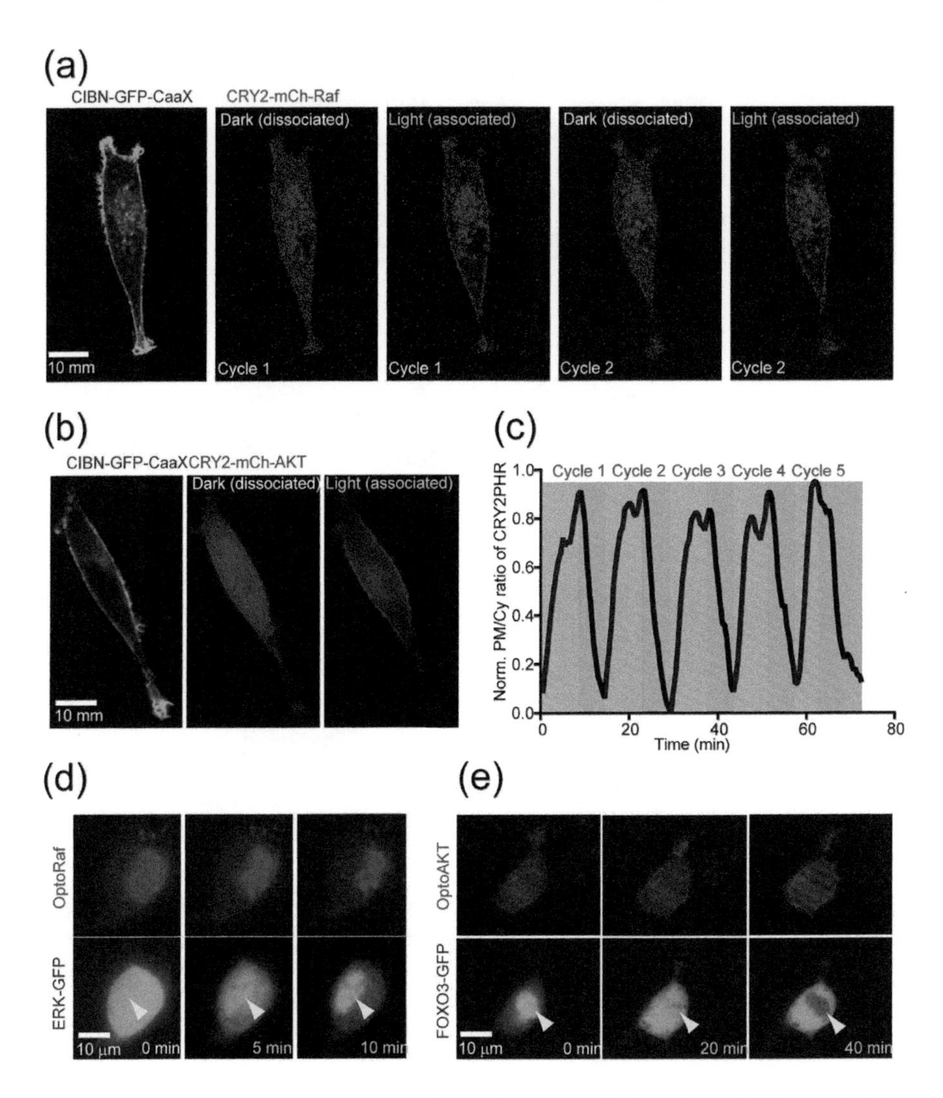

FIGURE 4.2 Reversible optogenetic stimulation of optoRaf and optoAKT revealed by live cell imaging. (a) Blue light induces membrane translocation of CRY2-mCh-Raf, which dissociates spontaneously in the dark. Multiple cycles of PM-to-cytosol shuttling can be achieved with repeated light-dark cycles. (b) Fusion of a signaling protein does not alter the reversible PM-to-cytoplasm shuttling demonstrated by the optoAKT system, which fuses AKT instead of Raf to CRY2-mCh as in (a). (c) Quantification of the association and dissociation kinetics of optoAKT during a course of 75 minutes of live cell stimulation. Membrane translocation of mCherry-fusion protein results in an increase of the fluorescence intensity ratio of the plasma membrane and the cytoplasm. (d) Activation of optoRaf causes ERK-GFP nuclear translocation resolved by live-cell imaging. (e) Activation of optoAKT causes FOXO3-GFP nuclear export resolved by live cell imaging. Scale bars: 10 μm. Panels adapted from [12] (CC BY 4.0).

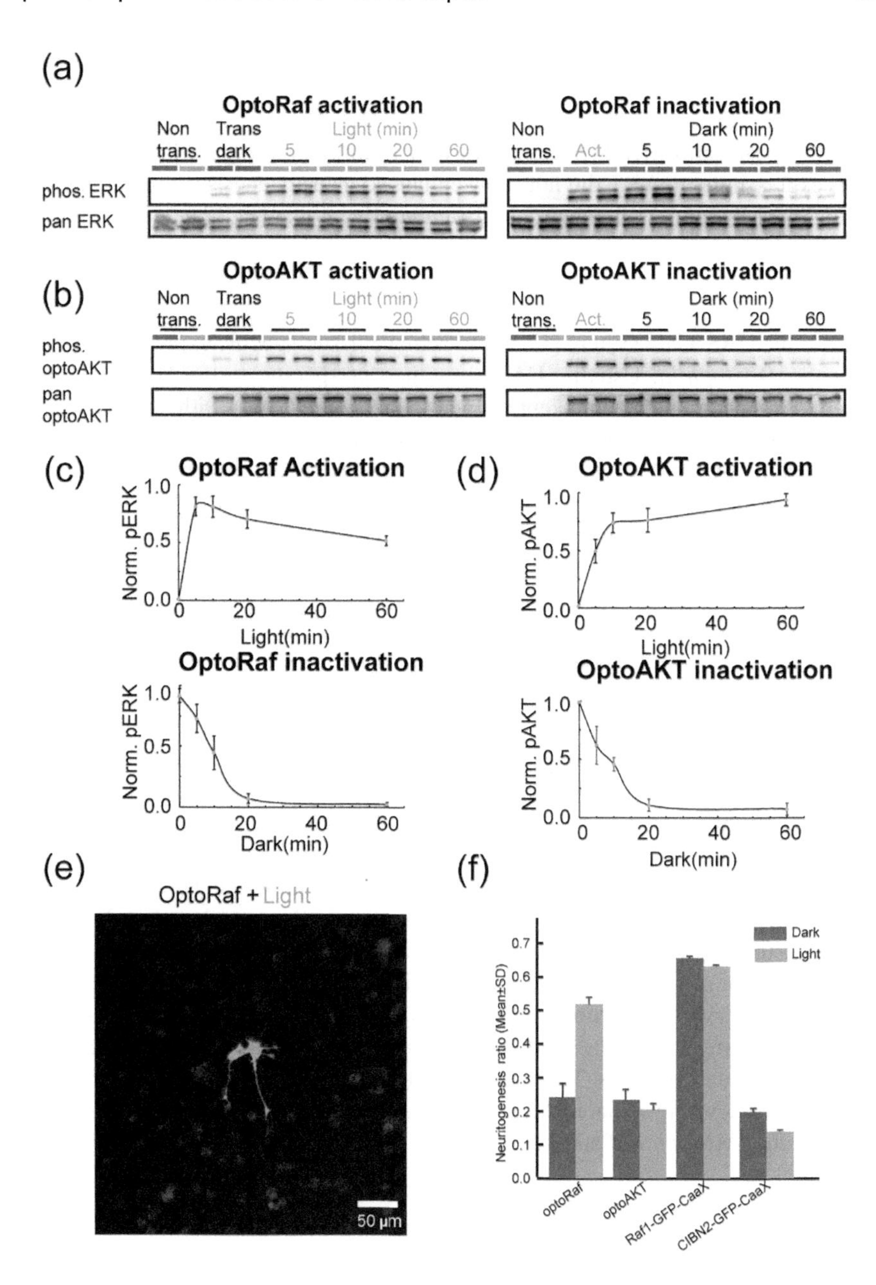

FIGURE 4.3 Biochemical and cellular responses induced by activation of optoRaf and optoAKT. HEK293 cells transfected with optoRaf (a) or optoAKT (b) show enhanced pERK and pAKT after blue light stimulation, as revealed by immunoblotting. Activation and inactivation kinetics of optoRaf (c) and optoAKT (d) were probed by Western blotting. (e) A representative fluorescence image of PC12 cells transfected with optoRaf undergoing neurite outgrowth 24 hours after blue light stimulation. Scale bars: 50 μm. (f) Quantification of neuritogenesis ratio of PC12 cells transfected with various plasmids under light and dark treatment. Panels adapted from [12] (CC BY 4.0).

To determine if optogenetic activation of Raf/MEK/ERK and AKT leads to functional outcomes, we carried out a cell neuritogenesis assay in PC12 cells transfected with optoRaf or optoAKT. Cells expressing optoRaf showed a significant increase of neuritogenesis ratio after 24 hours of blue light illumination (Figure 4.3e-f). However, cells expressing optoAKT did not show enhanced neuritogenesis (Figure 4.3f), indicating that ERK and AKT subcircuits differentially regulate PC12 neurite outgrowth.

4.4 ACTIVATION OF THE OPTOGENETIC SYSTEM IN FLIES

We use the UAS-Gal4 system [13, 14] to achieve tissue-specific expression of transgenes in *Drosophila*. Usually, UAS, an enhancer specific to the Gal4 protein, is fused to a gene of interest and kept in one fly line, while GAL4 with a tissue-specific promoter is kept in the other line. When these two lines are mated, the GAL4 protein will bind to UAS and activate gene transcription in specific tissues. We thus generated transgenic flies with inducible expression of optoRaf or optoAKT to examine the efficacy of the optogenetic systems *in vivo.*

4.4.1 Generation of the *UAS-optoRaf/AKT* Transgenic Flies

The pACU2 vector was modified from pUAST and is commonly used to generate transgenic flies in a site-directed manner [15, 16]. This vector contains a *Drosophila* P-element transposon, attB site for site-specific integration, mini-white (a reporter), and 5X UAS enhancer for high-level transgene expression. The pACU2 vector was digested with restriction enzymes NotI and XhoI to accommodate cDNA sequences encoding optoRaf or optoAKT. The constructed plasmid was injected into fly embryos by Rainbow Transgenic Flies, Inc. and the attB-containing transgenic vectors would be integrated into the attP40 docking site on the second chromosome at a certain probability [17]. The F0 injected flies were then incubated at 25°C until eclosion. To screen for those with stable optoRaf/AKT integration, we crossed every 2 injected F0 male flies to 5 w^{1118}; *Cyo/Sco* virgin females. In the F1 generation, the transgenic flies with the genome-integrated pACU2 vector expressed the mini-white reporter, making their eyes light yellow, thus distinguishing them from the non-transgenic white-eyed flies. The selected F1 transgenic males were again crossed to w^{1118}; *Cyo/Sco* virgin females. Male and female F2 transgenic flies were combined to make the *UAS-optoRaf* and *UAS-optoAKT* fly stocks.

4.4.2 Activation of optoAKT and optoRaf *in vivo*

The pickpocket (ppk) is a channel protein expressed explicitly in the class IV dendritic arborization (da) sensory neurons in flies, so ppk-Gal4 drives high expression of the transgene in class IV da (C4da) neurons [15, 18]. To test whether light stimulation would activate the Raf/MEK/ERK or AKT signaling in neurons, we crossed the *ppk-Gal4* flies *to UAS-optoRaf* or *UAS-optoAKT* flies, ensuring that expression of the transgene was restricted to C4da neurons. The offspring were kept in the dark for 72 hours, then wild-type (WT) and transgenic larvae were subjected to whole-field blue light illumination for 5, 10, and 15 min. For whole-field blue light stimulation, a 470 nm blue LED (LUXEON Rebel LED) was set on top of the fly incubator to expose the entire larva to light illumination.

After blue light illumination, larval body walls were dissected immediately as described [19] and incubated with the pERK1/2 antibody to probe Raf/MEK/ERK pathway activation; or with phospho-ribosomal S6 kinase ($p70^{S6K}$) downstream of AKT. We found that after 5-min light stimulation, the fluorescence intensity of pERK was increased in the neuron cell body, demonstrating that in optoRaf-expressing neurons, 5-min illumination was sufficient to activate the Raf/MEK/ERK signaling (Figure 4.4a-b). The intensity of pERK was further enhanced as the illuminating time increased. Moreover, after 15-min light treatment, ERK was significantly translocated into the nucleus. On the contrary, in larvae that were kept in the dark and not

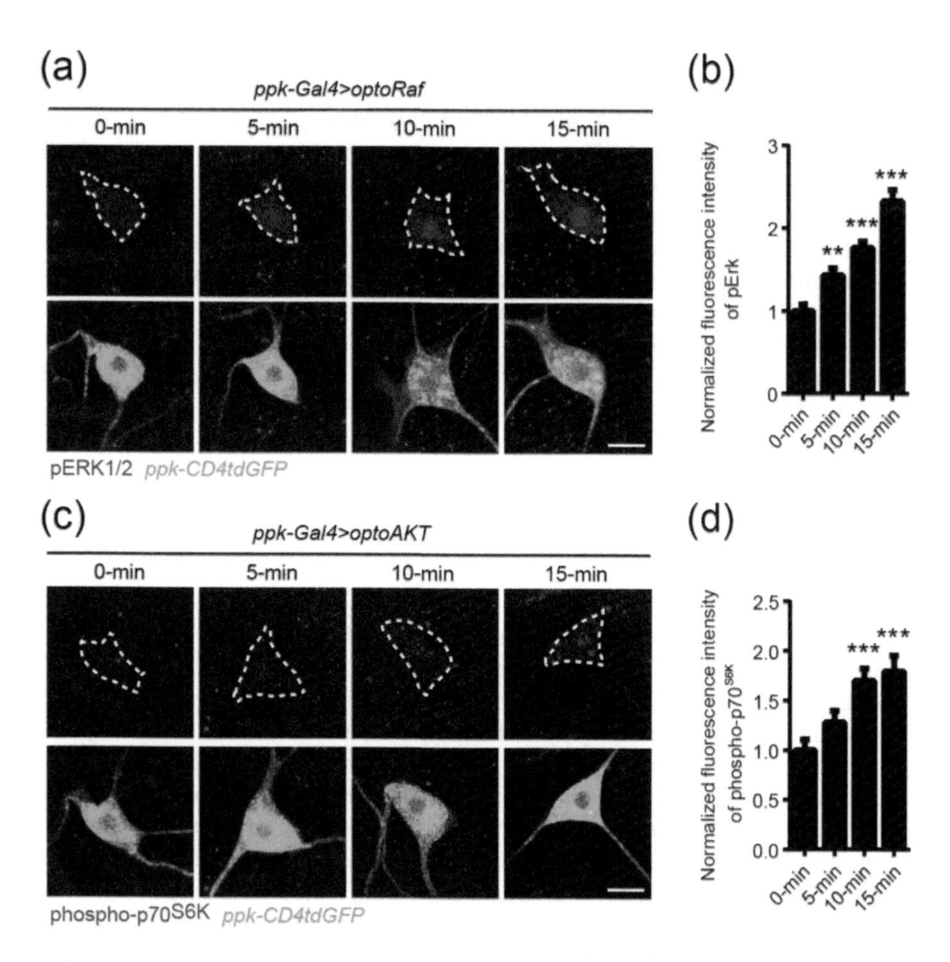

FIGURE 4.4 Activation kinetics of optoRaf and optoAKT in fly sensory neurons. (a) Five minutes of blue light illumination is sufficient to activate optoRaf, while longer light stimulation can further increase pERK intensity and induce ERK nuclear translocation. Scale bar: 10 μm. (b) The fluorescence intensity of pERK is normalized to that of neurons with no blue light treatment (0-min). (c) The intensity of phospho-$p70^{S6K}$ is significantly increased after 10-min light illumination in optoAKT expressing neurons. Scale bar: 10 μm. (d) The fluorescence intensity of phospho-$p70^{S6K}$ is normalized to that of neurons with no blue light treatment (0-min). Data are mean ± SEM, analyzed by one-way ANOVA followed by Dunnett's multiple comparisons test. **$P < 0.01$, ***$P < 0.001$. Panels adapted from [12] (CC BY 4.0).

exposed to blue light, the pERK signal was very weak and distributed evenly in the cytoplasm. In contrast, optoAKT was less sensitive to blue light. In neurons expressing optoAKT, the 5-min light illumination failed to substantially elevate the fluorescence intensity of phosphor-p70^{S6K}. But when the blue light stimulation increased to 10 or 15 min, the intensity was significantly higher than neurons kept in the dark (Figure 4.4c-d), suggesting that compared with optoRaf, a longer illuminating time is needed to activate the optoAKT signaling.

4.4.3 Inactivation of optoAKT and optoRaf

We also examined the reversibility of light-mediated Raf/MEK/ERK or AKT signaling activation. Larvae were treated with whole-field blue light for 15 min and then kept in the dark for 15 or 45 min. The pERK intensity slowly decreased after the light was shut off (Figure 4.5a), while the decay rate of phosphor-p70^{S6K} appeared to be much faster. In optoAKT expressing neurons, the intensity of phospho-p70^{S6K} decayed to the basal level after 15 min of dark incubation (Figure 4.5b). In contrast, in optoRaf expressing neurons, even when the off-time increased to 45 min, the intensity of pERK was still higher than neurons without light stimulation (Figure 4.5c-d). The difference in the inactivation kinetics reflected that optoRaf and optoAKT displayed different sensitivity in response to light when expressed in *Drosophila* C4da neurons. The optoAKT was less sensitive to light stimulation, and the AKT signaling was inactivated faster when the light was turned off. The data also showed that there was no crosstalk between optoRaf and optoAKT signaling (Figure 4.5a-b, **last column**).

4.5 LOCAL OPTOGENETIC STIMULATION TO GUIDE REGENERATING AXONS

In *Drosophila*, the axons of the ventral C4da neurons v'ada extend ventrally, showing a typical turn, then join the axon bundle (forming the axon-converging point) and project into the CNS. The C4da sensory neurons are known to regenerate robustly after axon injury in the peripheral nervous system (PNS) [20]. Nevertheless, most regenerating axons extend away from their original path. More than 60% of injured axons bifurcate and form two branches targeting opposite directions. In most cases, the dorsal branch, which extends towards the opposite trajectory, exhibits stronger growth competence than the ventral branch. For those that do not bifurcate, the axons usually regrow towards the dorsal trajectory. As a result, over 80% of regenerating axons grow away from the original ventral path [12]. This feature makes it an ideal model to test if spatially confined optogenetic activation of Raf/MEK/ERK or AKT pathways improves pathfinding.

4.5.1 Fly Sensory Neuron Injury Model

Previous reports suggest that axon regeneration programs are evolutionarily conserved [21]. Here, we used the *Drosophila* C4da sensory neuron injury model to explore whether we could take advantage of our optogenetic tools to manipulate the

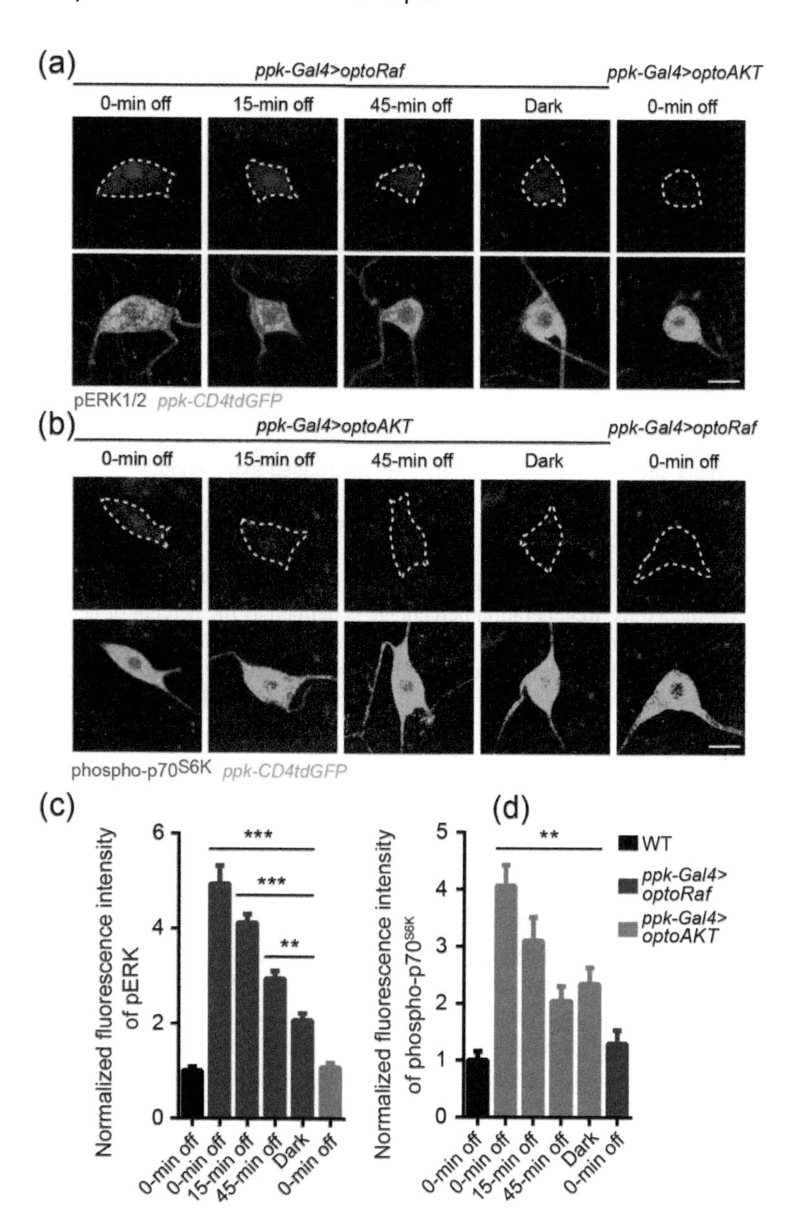

FIGURE 4.5 Inactivation kinetics of optoRaf and optoAKT in fly sensory neurons. (a) The body wall of larvae expressing optoRaf was dissected and stained for pERK1/2. The 15-min continuous light illumination leads to enhanced fluorescent intensity and nuclear translocation of pERK in the optoRaf-expressing C4da neurons (labeled by *ppk-CD4tdGFP*). The level of pERK decreases after the light is turned off but remains above the basal level 45 minutes in the dark. Notably, the ERK signaling is not activated by light stimulation in optoAKT-expressing neurons. (b) Inactivation kinetics of optoAKT probed by immunostaining of phospho-p70^{S6K}. OptoAKT is inactivated 15-min after the light is turned off. Note that activation of optoRaf does not upregulate phospho-p70^{S6K}. (c-d) Quantification of the inactivation kinetics in (a-b). Levels of fluorescence were normalized to the WT neurons. Panels adapted from [12] (CC BY 4.0).

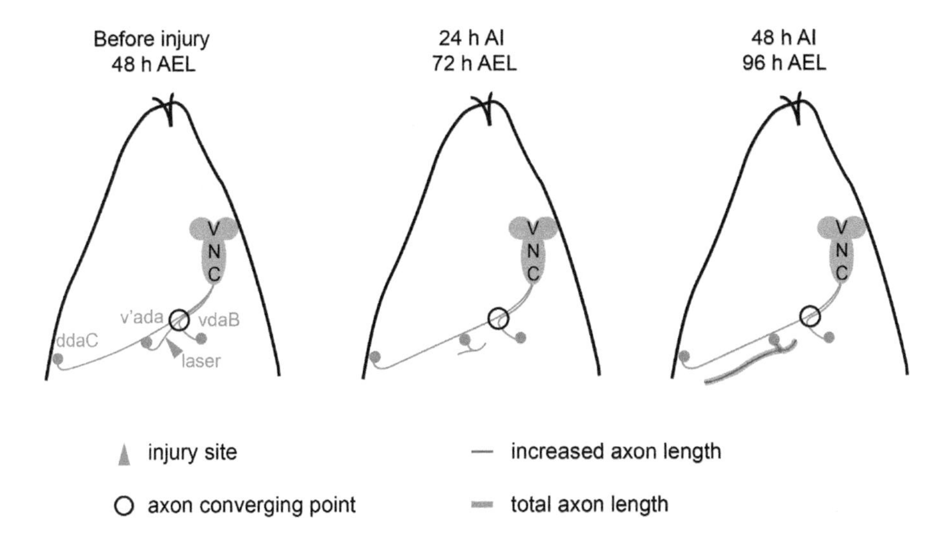

FIGURE 4.6 Scheme for a fly sensory neuron injury model. A 930 nm two-photon laser is used to sever the axons. Regeneration percentage is defined as the percent of neurons showing significantly elongated axons at 48 h AI compared with 24 h AI. Regeneration index is defined as the increased axon length normalized to the distance between the cell body and the axon-converging point. AI: after injury, h AEL hours after egg laying (h AEL).

neurotrophic signaling for guided axon regeneration. We crossed *UAS-optoRaf* or *UAS-optoAKT* flies with *ppk-Gal4; ppk-CD4-tdGFP* flies and kept them in the dark. Fly larvae were collected using grape juice agar plates. At 48 hours after egg laying (h AEL), we selected third instar larvae and anesthetized them with ether in a fume hood. Then we located the v'ada C4da neurons under the fluorescence microscope and used the "crop" function in the imaging software to focus the scan window onto the target axon (typically 40–50 μm from the cell body). We used a 930 nm two-photon laser to sever the axons with minimized photodamage. We set the gain to ~750 and laser intensity to ~600 mW, using a low scan speed (~4) under the continuous mode to scan the target area. After injuring the v'ada C4da neurons, the larvae were returned to grape juice agar plates. We performed confocal imaging of the injured neurons at 24 hours after injury (h AI) to ensure that axons were successfully severed and at 48 h AI to assess regeneration (Figure 4.6). We used "regeneration percentage" and "regeneration index" to quantify the extent of axon regeneration [20, 22]. Regeneration percentage is defined as the percent of neurons showing significantly elongated axons at 48 h AI compared with 24 h AI. Regeneration index is defined as the increased axon length normalized to the distance between the cell body and the axon-converging point.

4.5.2 Whole-Field Light Stimulation Promotes Axon Regeneration but Fails to Improve Pathfinding

Neurotrophins are well-documented as pro-growth factors to support axon regrowth after nerve injury [23]. The Raf/MEK/ERK and AKT signaling pathways are both implicated in regeneration. The AKT pathway was reported to promote axon regrowth in flies, and some studies showed that ERK is involved in axon extension [20, 24, 25]. First, we examined whether activation of neurotrophic subcircuits in the entire neuron could promote axon regeneration in optoRaf/optoAKT expressing C4da sensory neurons. The transgenic larvae were kept in the dark after egg laying. Axon injury was performed as described previously. After the injury, the larva was returned to a new grape juice agar plate and exposed to whole-field blue light throughout the remaining procedure. As expected, C4da neurons expressing optoRaf or optoAKT showed enhanced regeneration capacity in response to blue light, with a substantial increase in regeneration index, but there was no difference between WT and unstimulated transgenic flies (Figure 4.7a-c). We also tested the potential synergy between Raf/MEK/ERK and AKT signaling in regeneration. We co-expressed optoRaf and optoAKT in C4da neurons, but the activation of both Raf/MEK/ERK and AKT pathways in the same neuron failed to increase the regeneration index further (Figure 4.7b-c). Notably, the whole-field illumination did not improve the poor pathfinding of regenerating axons in the transgenic larvae.

4.5.3 Local Illumination of an Axon Branch Leads to Preferred Regeneration

A previous report revealed that activating the neurotrophic signaling in cell bodies and distal axons resulted in distinct cell responses, raising the possibility that local activation of Raf/ERK/MEK or AKT signaling might lead to different functional outcomes in injured neurons [6]. Since our optogenetic tools enable precise spatial control of the neurotrophic signaling in live animals, we thus investigated whether spatially restricted activation of the neurotrophic signaling could specifically enhance the regrowth of the ventral branch, which extends towards the correct trajectory.

The larva was kept in the dark throughout the whole procedure. After the first imaging time point at 24 h AI, we immediately used the confocal microscope to focus the blue light (delivered by the 488 nm argon laser) onto the entire ventral branch for 5 min. To weigh the regeneration of the ventral branch over the dorsal branch, we defined a new index, the relative regeneration ratio. We measured the lengths of both the ventral and dorsal branches at 24 and 48 h AI, and subtracted the increased dorsal branch length (Δdorsal) from the increased ventral branch length (Δventral), then divided the value by the total increased length of these two branches. The quotient was defined as the relative regeneration ratio. If the ventral branch exhibited more regenerative competence, the ratio would be positive; otherwise, it would be negative. Without light stimulation, the relative regeneration ratio of the transgenic larvae was comparable to that of WT, confirming the preferred regrowth of the dorsal branch (Figure 4.8a-b). Strikingly, the local blue light stimulation significantly increased the ratio in optoRaf- or optoAKT-expressing v'ada C4da neurons, while this transient stimulation resulted in no difference in WT (Figure 4.8a-b). In optoRaf expressing neurons, the mean value of relative regeneration ratio was positive, suggesting that the growth rate of the ventral branch surpassed the dorsal

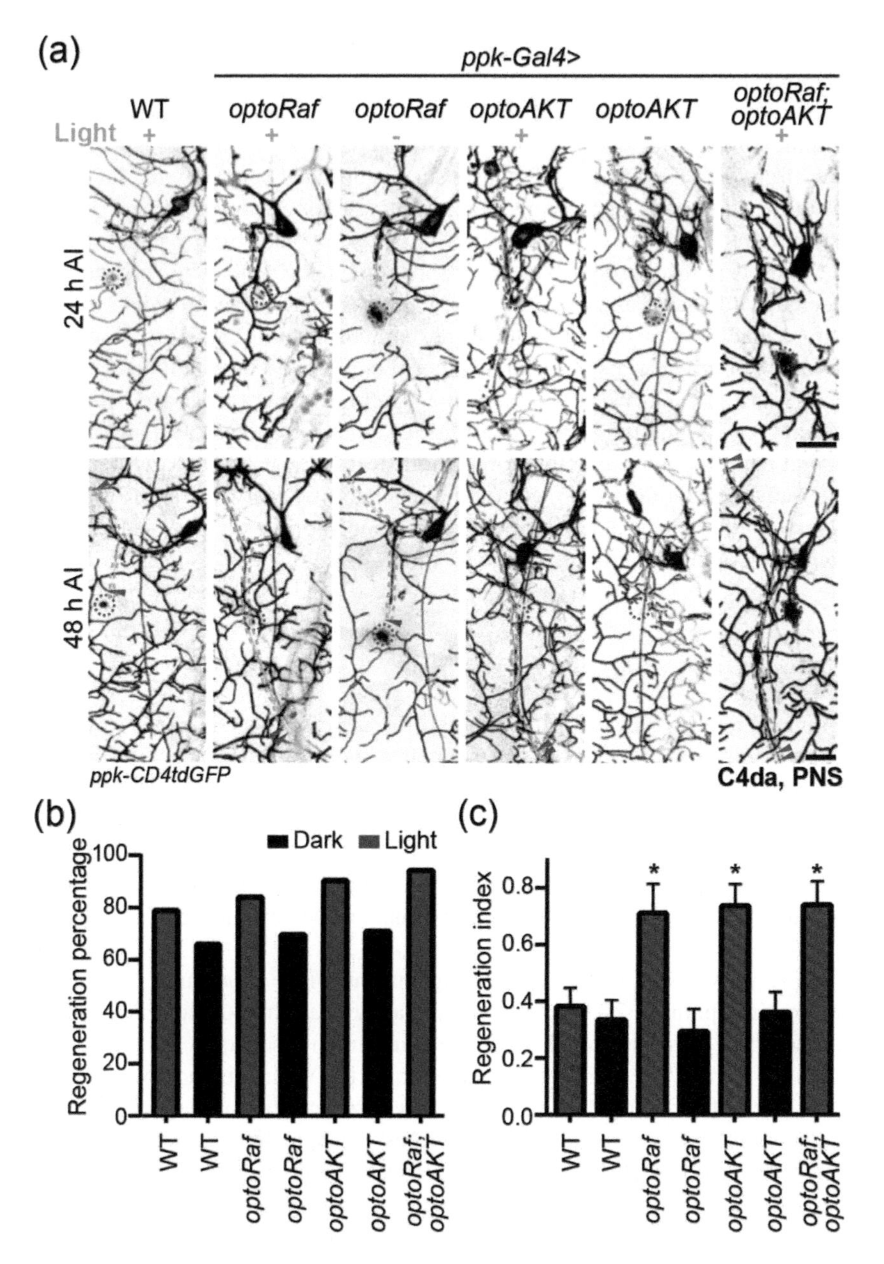

FIGURE 4.7 Light induces enhanced axon regeneration of C4da neurons in the PNS expressing optoRaf and optoAKT. (a) C4da neuron axons were severed and their regeneration was assayed at 48 h AI. The injury site is marked by the dashed circle and regenerating axons are marked by arrowheads. Axons are outlined with dashed green lines. Scale bar = 20 μm. (b) The regeneration percentage of light-stimulated transgenic groups is not significantly higher than WT. (c) Qualification of C4da neuron axon regeneration by the regeneration index. Transgenic groups with blue light stimulation show a higher regeneration index than WT. Panels adapted from [12] (CC BY 4.0).

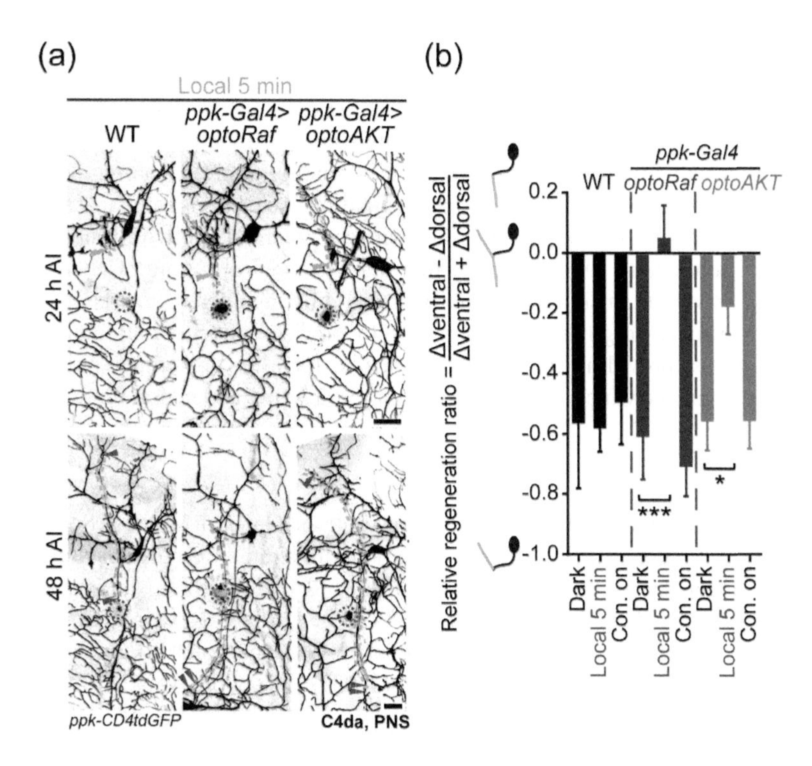

FIGURE 4.8 Spatial activation of optoRaf and optoAKT guides axon regeneration. A single pulse of light stimulation delivered specifically on the ventral axon branch at 24 h AI (blue flash symbol) is capable of promoting the preferential extension of regenerating axons in optoRaf- or optoAKT-expressing larvae. (a) Qualification of the relative regeneration ratio of C4da neuron of PNS. Data are shown as mean ± SEM, analyzed by one-way ANOVA followed by the Dunnett's multiple comparisons test. * $P < 0.05$, *** $P < 0.001$. Panels adapted from [12] (CC BY 4.0).

branch. This result suggests that Raf/MEK/ERK signaling may be more effective in promoting axon elongation.

As mentioned, whole-field light illumination failed to increase the relative regeneration ratio in transgenic larvae. This result thus reveals that, after injury, activating the neurotrophic signaling in a single axon branch may induce different biological processes from activating the signaling in the entire neuron. Taken together, our data demonstrate that transient local stimulation affects the decision-making of the growth cone at the branching point, but in order to increase overall axon regeneration, chronic stimulation of the neurotrophic signaling is necessary.

4.6 CONCLUDING REMARKS AND PERSPECTIVE

Functional outcomes of many signaling pathways depend on not only the activity of each signaling component but also on their dynamic interactions in time and space. Experimental strategies that allow for the direct probing of signaling dynamics in live cells remain limited. Opsin-free optogenetics modulates molecular activity in live cells dynamically and expands its target into a diverse pool of signaling enzymatic molecules. This modality makes optogenetics an attractive strategy to delineate signaling mechanisms within subcellular compartments, which promises to offer new insights into the fundamental understanding of cell signaling. Using our optogenetic system based on neurotrophin signaling, we demonstrated its capacity to promote robost axon regeneration *in vivo*, with the potential to circumvent the undesired pain subcircuits. Our system further allows temporal tuning of neurotrophin signaling activation, raising the possibility of modulating axon regeneration dynamics. We also achieved guided axon regrowth by precise light stimulation on specific axon branches, providing the proof-of-principle for allowing proper guidance of regenerating axons, which is critical for axon re-targeting and neural circuit repair.

ACKNOWLEDGMENTS

This work is supported by the National Institute of General Medical Sciences of the National Institutes of Health and the National Institute of Environmental Health Sciences under Award Number R01GM132438, R01MH124827, National Science Foundation under Award Number 2121003, and Scialog®, Research Corporation for Science Advancement (RCSA), the Frederick Gardner Cottrell Foundation, and the Paul G. Allen Frontiers Group (Award #27937) (K.Z.); the National Institute of Neurological Disorders and Stroke of the National Institutes of Health under Award Number 1R01NS107392 (Y.S.).

REFERENCES

1. Armour, B.S., et al., Prevalence and causes of paralysis-United States, *2013. Am J Public Health*, 2016. **106**(10): pp. 1855–1857.
2. Ramer, M.S., J.V. Priestley, and S.B. McMahon, Functional regeneration of sensory axons into the adult spinal cord. *Nature*, 2000. **403**(6767).
3. Park, K.K., et al., Promoting axon regeneration in the adult CNS by modulation of the PTEN/mTOR pathway. *Science*, 2008. **322**(5903): pp. 963–966.
4. Mahar, M. and V. Cavalli, Intrinsic mechanisms of neuronal axon regeneration. *Nat Rev Neurosci*, 2018. **19**(6): pp. 323–337.
5. Marshall, C.J., Specificity of receptor tyrosine kinase signaling: Transient versus sustained extracellular signal-regulated kinase activation. *Cell*, 1995. **80**(2): pp. 179–185.
6. Watson, F.L., et al., Neurotrophins use the Erk5 pathway to mediate a retrograde survival response. *Nat Neurosci*, 2001. **4**(10): pp. 981–988.
7. Oh, T.J., et al., Steering molecular activity with optogenetics: Recent advances and perspectives. *Adv Biol (Weinh)*, 2021. **5**(5): p. e2000180.
8. Ong, Q., et al., The timing of Raf/ERK and AKT activation in protecting PC12 cells against Oxidative Stress. *PLoS One*, 2016. **11**(4): p. e0153487.
9. Zhang, K., et al., Light-mediated kinetic control reveals the temporal effect of the Raf/MEK/ERK pathway in PC12 cell neurite outgrowth. *PLoS One*, 2014. **9**(3): p. e92917.
10. Kennedy, M.J., et al., Rapid blue-light—mediated induction of protein interactions in living cells. *Nature Mathods*, 2010. **7**(12): pp. 973–977.
11. Krishnamurthy, V.V., et al., Reversible optogenetic control of kinase activity during differentiation and embryonic development. *Development*, 2016. **143**(21): pp. 4085–4094.

12. Wang, Q., et al., Optical control of ERK and AKT signaling promotes axon regeneration and functional recovery of PNS and CNS in Drosophila. *Elife*, 2020. **9**.
13. Liu, Y. and M. Lehmann, A genomic response to the yeast transcription factor GAL4 in Drosophila. *Fly (Austin)*, 2008. **2**(2): pp. 92–98.
14. Webster, N., et al., The yeast UASG is a transcriptional enhancer in human HeLa cells in the presence of the GAL4 trans-activator. *Cell*, 1988. **52**(2): pp. 169–178.
15. Han, C., L.Y. Jan, and Y.N. Jan, Enhancer-driven membrane markers for analysis of nonautonomous mechanisms reveal neuron-glia interactions in Drosophila. *Proc Natl Acad Sci U S A*, 2011. **108**(23): pp. 9673–9678.
16. Makridou, P., et al., Hygromycin B-selected cell lines from GAL4-regulated pUAST constructs. *Genesis*, 2003. **36**(2): pp. 83–87.
17. Venken, K.J., et al., Versatile P[acman] BAC libraries for transgenesis studies in Drosophila melanogaster. *Nat Methods*, 2009. **6**(6): pp. 431–434.
18. Grueber, W.B., et al., Dendrites of distinct classes of Drosophila sensory neurons show different capacities for homotypic repulsion. *Curr Biol*, 2003. **13**(8): pp. 618–626.
19. Brent, J.R., K.M. Werner, and B.D. McCabe, Drosophila larval NMJ dissection. *J Vis Exp*, 2009(24).
20. Song, Y., et al., Regeneration of Drosophila sensory neuron axons and dendrites is regulated by the Akt pathway involving Pten and microRNA bantam. *Genes Dev*, 2012. **26**(14): pp. 1612–1625.
21. Fang, Y. and N.M. Bonini, Axon degeneration and regeneration: Insights from Drosophila models of nerve injury. *Annu Rev Cell Dev Biol*, 2012. **28**: pp. 575–597.
22. Song, Y., et al., Regulation of axon regeneration by the RNA repair and splicing pathway. *Nat Neurosci*, 2015. **18**(6): pp. 817–825.
23. Ramer, M.S., J.V. Priestley, and S.B. McMahon, Functional regeneration of sensory axons into the adult spinal cord. *Nature*, 2000. **403**(6767): pp. 312–316.
24. Huang, H., et al., PI3K/Akt and ERK/MAPK signaling promote different aspects of neuron survival and axonal regrowth following rat facial nerve axotomy. *Neurochem Res*, 2017. **42**(12): pp. 3515–3524.
25. Markus, A., J. Zhong, and W.D. Snider, Raf and akt mediate distinct aspects of sensory axon growth. *Neuron*, 2002. **35**(1): pp. 65–76.

5 Optogenetic Control of Neural Stem Cell Differentiation

Yixun Su, Taida Huang, Kai Zhang, and Chenju Yi

CONTENTS

5.1 Introduction 71
5.2 Application of Optogenetic Tools in the Nervous System *in vitro* 72
5.3 Optogenetic Control of Mouse Neural Progenitors via OptoRaf1 73
5.3.1 Cell Culture 73
5.3.2 Lentivirus Plasmid Construction 74
5.3.3 Lentivirus Preparation and Neural Progenitor Infection 74
5.3.4 Light Stimulation 74
5.3.5 Sample Collection and Analysis 75
5.3.6 Results and Perspectives 76
5.4 Optogenetic Control of Hair Follicle-Derived Stem Cell Differentiation via OptoTrkA 79
5.4.1 Primary Culture of Hair Follicle-Derived Stem Cells 79
5.4.2 Plasmid Construction 79
5.4.3 Cell Transfection 79
5.4.4 Light Stimulation 80
5.4.5 Sample Collection and Analysis 81
5.4.6 Results and Perspectives 82
5.5 Concluding Remarks 82
Acknowledgments 82
References 82

5.1 INTRODUCTION

Cells are constantly sensing and responding to extracellular stimuli in the environment, which is manifested as activation or inactivation of different intracellular signaling pathways, in which receptors (or receptor complexes) receive the stimuli and activate a cascade of signaling proteins to regulate cellular processes such as transcription, cell division, or migration, to name a few. Intriguingly, distinct stimuli (growth factors, for example) could utilize the same set of intracellular signaling pathways yet cause different outputs [1]. The specificity of output could be achieved by the spatiotemporal regulation of signaling components. However, conventionally

DOI: 10.1201/b22823-5

used pharmacological and genetic approaches could not attain sufficient spatiotemporal resolution to dissect the dynamic information in intracellular signal transduction.

Optogenetic tools utilize light to manipulate protein-protein interactions, which are initially applied in the regulation of neuronal activity based on opsins [2, 3]. After years of development, opsin-free optogenetic toolboxes have been expanded to include light-sensitive protein pairs with different photo-sensing properties and association/dissociation kinetics [4]. Opsin-free optogenetics offers a method to control the activation or inactivation of signaling components precisely and flexibly in live cells [5].

5.2 APPLICATION OF OPTOGENETIC TOOLS IN THE NERVOUS SYSTEM *IN VITRO*

The nervous system development relies heavily on cell-cell or cell-autonomous signaling. For instance, neural progenitors produce neurons at the early embryonic stage but switch to a gliogenic fate at the later stages, differentiating into astrocytes and oligodendrocyte precursors [6, 7]. Multiple signaling pathways are involved in this fate transition, including the RAS/RAF/MEK/ERK pathway [6][8]. Upon the stimulation of growth factors such as fibroblast growth factors (FGFs) or ciliary neurotrophic factors (CNTFs), their receptors dimerize and recruit adapter proteins that activate membrane-bound small GTPase RAS, which recruits RAF protein to the plasma membrane, where it is phosphorylated and further activates downstream kinase cascade MEK and ERK1/2. Active ERK1/2 translocates into the nucleus to regulate gene transcription [9]. Elements of the RAS/RAF/MEK/ERK pathway have been shown to promote the maintenance of neural progenitor state and their differentiation into astrocytes [10–13]. For instance, gain-of-function RAF mutants cause astrocyte overproduction but do not alter the neuron number [11, 12]. Constitutive activation of MEK leads to excessive astrocytogenesis and reduced neurogenesis [14]. However, other reports showed that RAF downstream pathway is necessary for neurogenesis and neurite outgrowth [15–17] and that the MEK-C/EBP axis promotes cortical neurogenesis [18]. The discrepancy in these observations could arise from the different time and kinetics of RAF activation [10]. Therefore, whether the temporal kinetics of the RAF/MAK/ERK signaling could cause different outcomes during neural progenitor differentiation remains to be elucidated. To tackle this problem, we developed an optogenetic system named OptoRaf1 based on the blue-light-sensitive *Arabidopsis thaliana* cryptochrome 2 (CRY2) and the N-terminal domain of cryptochrome-interacting basic-helix-loop-helix 1 (CIBN). CIBN is anchored to the plasma membrane via a RAS CaaX domain and RAF1 is fused to CRY2 protein. Blue light stimulation causes CIBN-CRY2 dimerization, which enables RAF1 membrane translocation [19, 20]. As demonstrated subsequently, reversible activation of OptoRaf1 allows us to dissect how timed activation of RAF1 regulates astrocytogenesis.

Notably, trans-differentiation of other stem cells into the neural cell is also regulated by numerous signaling pathways. Understanding the role of these pathways in trans-differentiation would be beneficial for the translational studies of stem cell therapy. As safety is one of the most concerning issues, there are advantages of using adult stem cells to develop cell-based transplantation therapy. For example, c-Myc used for creating iPSC from the somatic cells is a notorious oncogene [21, 22]. In addition, the protracted *in vitro* culture of both induced pluripotent stem cells (iPSCs) and embryonic stem cells (ESCs) could induce pro-oncogenic genomic aberrations [23]. Hair follicle-derived stem cells (HSC) are an ideal source for neural regeneration because of their easy access, high differentiation potency, and low tumorigenicity [24–26]. It has been suggested that neurotrophic pathways play an essential role in HSC differentiation. Neurotrophin ligand binds to the tropomyosin receptor kinase (TrkA) family and, alternatively, the p75 neurotrophin receptor, activating downstream pathways, including the ERK and AKT pathways that are key to neural stem cell survival, proliferation, and differentiation [27, 28]. However, the role of TrkA-mediated signaling in the HSC trans-differentiation remains to be elucidated. Thus, we developed another optogenetic system called OptoTrkA that expresses a Lyn-AuLOV-TrkA(ICD)-GFP fusion protein. The membrane-anchored TrkA intracellular domain could form homodimers via interaction of AuLOV upon 450 nm-light stimulation [29]. Compared with genetic overexpression and pharmacological stimulation, light-induced activation of OptoTrkA is blessed with spatiotemporal accuracy and minimal off-target side effects.

In this chapter, we showcase applications of optogenetic tools in nervous system development, stem cell trans-differentiation, HSC proliferation, migration, and neural differentiation.

5.3 OPTOGENETIC CONTROL OF MOUSE NEURAL PROGENITORS VIA OPTORAF1

5.3.1 Cell Culture

Mouse neural progenitors can be isolated from embryonic day 13~15 (E13~E15) mouse brain. Neural progenitors isolated from the later stages have a higher gliogenic tendency. Note that the animal-related protocols must be approved by the Institutional Animal Care and Use Committee. At the desired embryonic stage (e.g., E15), the dam is euthanized by CO_2 overdose, followed by dissection of the uterus. On ice, the uterus is placed in a 10 cm petri dish filled with Dulbecco's phosphate buffered saline (DPBS), and embryos are isolated and transferred to a new 10 cm petri dish with DPBS. The embryonic mouse brains are carefully dissected, meninges removed, and transferred to a 15 mL falcon tube with DPBS. After all the brains are isolated, the DPBS in the falcon tube is then replaced by ~ 2 mL Accutase (Thermo Fisher), which is then gently rocked for 15 min in a 37°C incubator for enzymatic digestion. The tube is then centrifuged at 1,000 g for 5 min, and the Accutase is replaced by 10 mL culture medium (Dulbecco's Modified Eagle Medium (DMEM)/F12 supplemented with B27, 0.6% D-glucose, 5 mM HEPES, 62.5 ng/mL progesterone, 100 nM putrescine, 1.83 μg/mL heparin, 20 ng/mL epidermal growth factor (EGF), 5 ng/mL basic FGF), followed by trituration for 20 times to dissociate neural progenitors. After cell counting, cells are diluted to 2×10^5 cells/mL and maintained in culture medium for 4 days as suspension culture to expand the population. Meanwhile, the cells would grow into spheres termed neurospheres. To allow cell differentiation, the neurospheres are dissociated by Accutase digestion, and the neural progenitors are seeded onto poly-D-lysine coated surface at 2×10^4 cells/cm^2, followed by growth factor withdrawal the next day.

5.3.2 Lentivirus Plasmid Construction

Because the primary neural progenitors are difficult to transfect using commercial transfection reagents, lentivirus is employed to deliver OptoRaf1 into the cells. The OptoRaf1-expressing lentiviral plasmid is constructed to express the CRY-mCherry-humanRAF1 and the CIBN-GFP-CAAX fusion protein. The original bicistronic OptoRaf1 based on a P2A self-cleaving peptide results in a large (~10 kb) viral genome challenging for efficient lentivirus packaging. Thus, we clone two fusion proteins separately into the pLJM-eGFP vector by substituting the eGFP fragment using seamless cloning enzymes (Figure 5.1).

5.3.3 Lentivirus Preparation and Neural Progenitor Infection

HEK293T cells used to produce lentivirus are maintained in high glucose DMEM supplemented with 10% fetal bovine serum (FBS) and 1% penicillin-streptomycin, and are subcultured every 3 days. The lentivirus is produced using the 3rd generation packaging systems. The envelope plasmid, the packaging plasmid, and the pLJM1-eGFP/pLJM1-CRY-mCherry-RAF1/pLJM1-CIBN-eGFP-CAAX plasmids are transfected into HEK293T cells using Lipofectamine 3000 when reaching ~70% confluency. The culture medium is renewed 6 hours after transfection. The lentivirus-containing media is collected twice at 24 hours and 48 hours after transfection and subjected to virus concentration by centrifugation in a 100 kDa cutoff ultracentrifuge tube at 1,500 g for 20 min. The resulting virus particles are tested for titers and stored at –80°C.

To infect neural progenitors, the neurospheres at day 4 *in vitro* are dissociated using Accutase digestion and resuspended in the culture medium. The lentivirus particles are added to the cell suspension. Cells are then seeded in poly-D-lysine coated cell culture plates with or without coverslips. The infected cells should avoid exposure to room light that could activate the OptoRaf1. To verify successful lentivirus transduction, fluorescence imaging (GFP channel) is performed one day after the infection. Expression of GFP in cells indicates a successful transfection.

5.3.4 Light Stimulation

A blue LED light source (~450 nm) is used to activate the OptoRaf1. The light intensity is adjusted to 25 LUX at the cell level to avoid phototoxicity. An intermittent 20 min on/40 min off cycle is used to achieve optimal OptoRaf1 activation. Note that different on/off cycles should be tested when a new cell type is used. For a prolonged period of light stimulation, a cooling device (e.g., a fan) can be used to avoid overheating the cells.

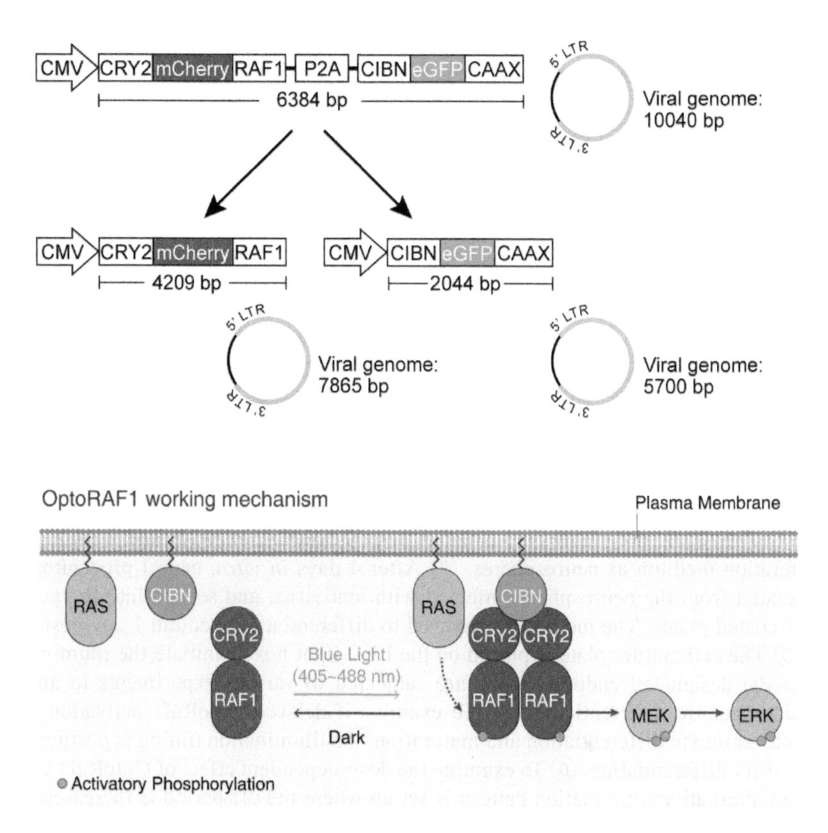

FIGURE 5.1 Scheme of lentiviral transfer plasmids and working mechanism of OptoRaf1. Up: Expression of two fusion proteins in a single transgene results in a large viral genome (>10,000 bp) that significantly lowers the virus titer. Thus, the two fusion proteins coding sequences are cloned into two different constructs for lentivirus production. Below: the working mechanism of the OptoRaf1. The blue light illumination triggers the interaction between membrane-anchored CIBN and CRY2, thereby recruiting CRY2-RAF1 fusion protein to the membrane. This allows the activation of RAF1 by membrane-located RAS proteins, leading to activation of the RAF/MEK/ERK pathway.

5.3.5 Sample Collection and Analysis

After light illumination, immunostaining is performed for cells cultured on the coverslips. The medium is removed from the well, followed by a rinse with ice-cold PBS, and the cells are fixed by 4% paraformaldehyde (PFA) at room temperature for 10 min. The cells are then rinsed thrice with PBS and subjected to immunostaining. For western blot analysis, cells are directly lysed with RIPA buffer supplemented with protease inhibitors after medium removal and PBS rinsing. For RT-qPCR analysis, cells are lysed by RNAzol and subjected to RNA isolation. For flow cytometry analysis, cells are dissociated by Accutase digestion, followed by fixation in 2% PFA and permeabilization in 90% methanol pre-cooled at –80°C, before being subjected to intracellular epitope staining. The workflow of our experiment is illustrated in Figure 5.2.

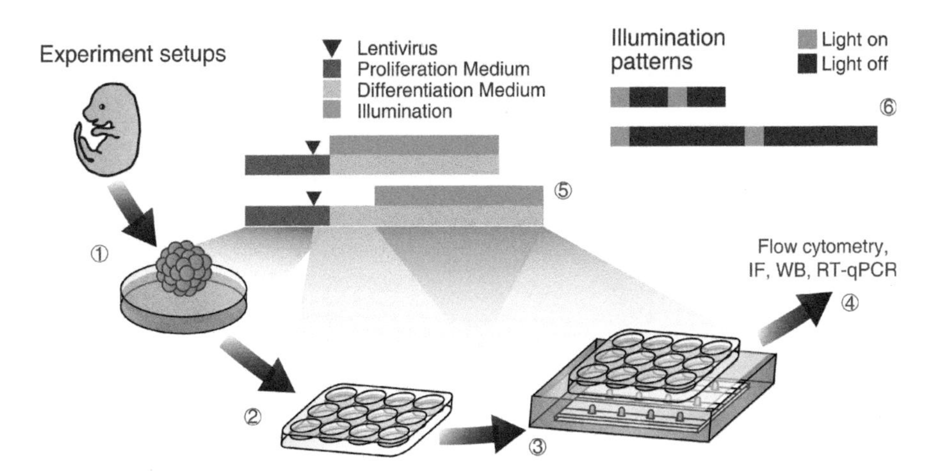

FIGURE 5.2 The workflow of the application of OptoRaf1 tool in mouse neural progenitors. ① The neural progenitors are isolated from E15 mouse embryos and maintained in proliferation medium as neurospheres. ② After 4 days *in vitro*, neural progenitors are dissociated from the neurosphere, infected with lentivirus, and seeded into the poly-D-lysine-coated plates. The medium is changed to differentiation medium 1 day post seeding. ③ The cell culture plate is placed on the LED light box to initiate the illumination. ④ At the designated endpoint, cells are subjected to various experiments to analyze neural progenitor differentiation. ⑤ To examine if delayed OptoRaf1 activation could promote astrocyte differentiation and maturation, the illumination timing is postponed to 2 days after differentiation. ⑥ To examine the dose-dependent effect of OptoRaf1 activation, an alternative illumination pattern is set up where the off period is increased from 40 min to 120 min.

5.3.6 Results and Perspectives

In this experiment, we set up 4 different experiment groups, including a GFP-expressed, maintained in the dark control group, an OptoRaf1-expressed, dark group (to verify the basal activity of optoRaf1), a GFP-expressed, light-illuminated group (to examine phototoxicity), and finally, the OptoRaf1-expressed, light-illuminated target group. The expression of glial fibrillary acidic protein (GFAP) and connexin43 (Cx43) is analyzed to determine the astrocyte differentiation and maturation, respectively [7, 30].

Flow cytometry analysis verifies that upon light stimulation, OptoRaf1 induces a higher phosphorylation level of ERK1/2, a downstream target of RAF1 pathway. Immunostaining result reveals that light stimulated OptoRaf1 activation yields a significant increase in the percentage of GFAP+ astrocytes, indicating an enhanced astrocytogenesis. Meanwhile, the expression of Cx43, a channel protein expressed by astrocytes that marks that astrocyte maturity, is significantly increased in the light-illuminated, OptoRaf1-expressed groups. Neither the OptoRaf1-expressed, dark group nor the GFP-expressed, light-illuminated group displays these phenotypes, suggesting that the light-illumination or the OptoRaf1 expression alone could not promote astrocytogenesis or astrocyte maturation (Figure 5.3).

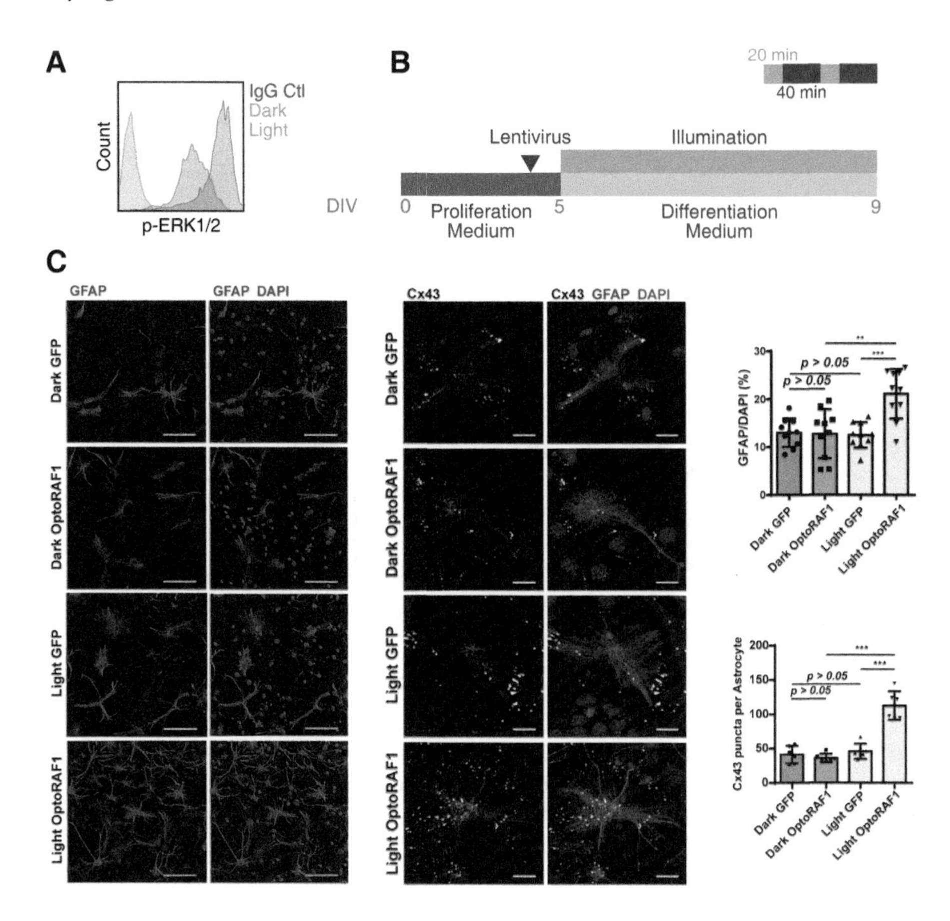

FIGURE 5.3 Activation of OptoRaf1 promotes astrocyte differentiation of neural progenitors. (a) Flow cytometry result shows that illumination promoted ERK1/2 phosphorylation in OptoRaf1 expressed neural progenitors, confirming the activation of the RAF/MEK/ERK pathway. (b) Experimental setup. (c) Immunofluorescence experiments show that light-induced OptoRaf1 activation (Light OptoRaf1 group) led to increase of the percentage of GFAP+ astrocytes and the expression of Cx43.

In order to observe the effect of time-dependent or dose-dependent manipulation of the OptoRaf1, we set up two additional experiments. Firstly, we delay the illumination window to two days post differentiation induction. The delayed OptoRaf1 activation fails to induce an increased GFAP+ astrocyte differentiation. However, it is still sufficient to promote Cx43 expression (Figure 5.4). Secondly, we change the illumination pattern to 20 min on/120 min off. The elongation of the dark period reduces the OptoRaf1 activation level, which fails to promote the expression of GFAP but could still promote Cx43 expression (Figure 5.5). It would be interesting in the future to apply this tool *in vivo* for a better understanding of the role of RAF1 in astrocytogenesis and astrocyte maturation.

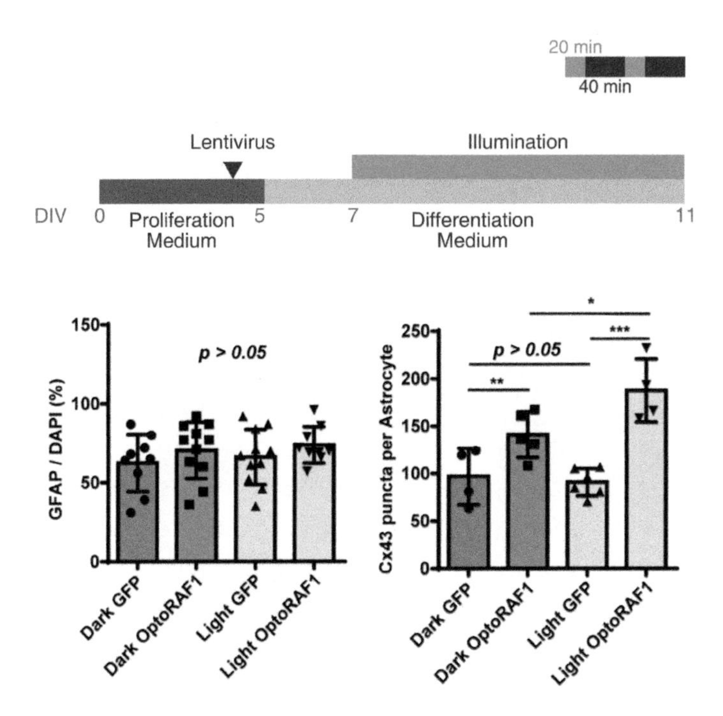

FIGURE 5.4 Delayed OptoRaf1 activation did not alter astrocytogenesis. Illumination timing is delayed to two days post differentiation induction. Immunofluorescence result shows that delayed OptoRaf1 activation do not alter the percentage of GFAP+ astrocyte; however, the Cx43 expression is still upregulated.

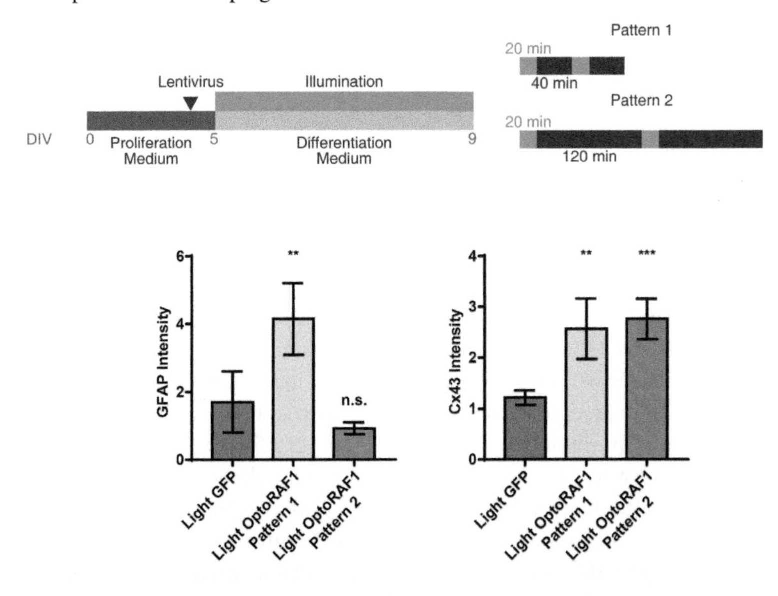

FIGURE 5.5 Dose-dependent effect of OptoRaf1 activation on astrocytogenesis. In the illumination pattern 2, the dark period is elongated to 120 min. The elongation of the dark period reduces the overall OptoRaf1 activation level, fails to promote the expression of GFAP, but could still promote Cx43 expression.

5.4 OPTOGENETIC CONTROL OF HAIR FOLLICLE-DERIVED STEM CELL DIFFERENTIATION VIA OPTOTRKA

5.4.1 Primary Culture of Hair Follicle-Derived Stem Cells

All the animal-related protocols must be approved by the Institutional Animal Care and Use Committee. Before microdissection, all the metal surgical tools, such as microblade and forceps are dry heated at 180°C for at least 5 hours for sterilization. The postnatal day 9–14 (P9–14) mice are humanely killed by cervical dislocation. The facial area of the sacrificed mouse is soaked to sterilize in a mixed solution consisting of 3% medical hydrogen peroxide product, and 1% medical povidone iodine solution. After 2 min, the remaining solution should be washed off with 75% alcohol. Under a stereo microscope, the whisker pads are dissected out and immersed in DPBS in a Petri-dish by removing fat or dermal tissues adhering to the inner side of the whisker pads to expose the vibrissa hair follicles. The whisker pads are sheared into strips following the arrangement of hair follicles, thus isolating strips containing several columns of hair follicles. Next, the coherent epidermis between columns of hair follicles is snipped. For further isolation, two small incisions are made on the upper and lower regions of the connective tissue capsule to release blood, and the sebaceous gland and the hair dermal papilla at two terminals are removed. Then a longitudinal incision along the hair follicle is made to open the connective tissue capsule surrounding the bulge. The bulge can be rolled out from the connective tissue capsule following the previous longitudinal incision. The isolated bulges are placed in a 4-well culture dish with collagen type I coated surface. The Minimum Essential Medium α (α-MEM) medium containing 10% FBS is added to each well, and the explants are cultured in a 37°C, 5% CO_2 incubator. After 3 days in culture, the hair follicle explants are removed and emigrated HSCs are further cultured in cell expansion medium which contains DMEM/F12, 10% FBS, 1 × B27 supplement, 10 ng/mL FGF, 10 ng/mL EGF, 50 ng/mL glial-cell-derived neurotrophic factor (GDNF). Primary cultured HSCs are dissociated and passaged using Accutase on day 7 (Figure 5.6).

5.4.2 Plasmid Construction

The plasmid expressing Lyn-TrkAICD-AuLOV-GFP is constructed by fusing fragments coding the Lyn lipidation tag, the rat TrkA ICD and the AuLOV into a pEGFP-N1 vector (Figure 5.7). For midi-prep, plasmids are extracted from 200 mL of overnight LB broth culture by using a commercialized midi-prep kit. The quality of plasmids is confirmed by the OD260/280 ratio measured with a spectrophotometer.

5.4.3 Cell Transfection

Cell transfection of HSC is performed with LipoStem Reagent (Invitrogen) by following the manufacturer's instruction. To obtain the optimal efficiency in cell transfection, 100,000 HSCs are seeded in a 24-well plate overnight before transfection. After 24 hours, the culture medium is renewed and the GFP signaling can be observed under an inverted fluorescent microscope.

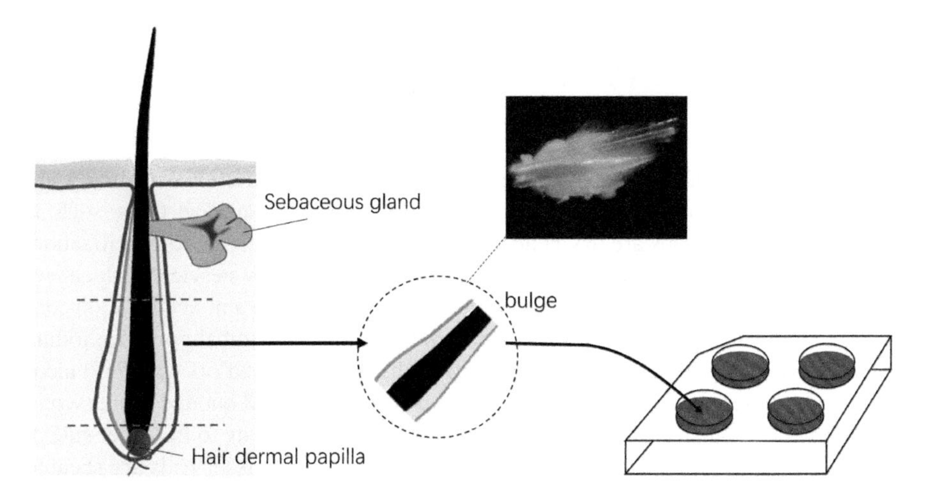

FIGURE 5.6 Schematic image showing the explant preparation for HSC primary culture. To isolate the bulge of hair follicle, two small incisions (red dot lines) are made on the upper and lower regions of the connective tissue capsule to remove the sebaceous gland and the hair dermal papilla. The isolated bulge explant is seeded onto the collagen-coated surface in a 4-well culture dish.

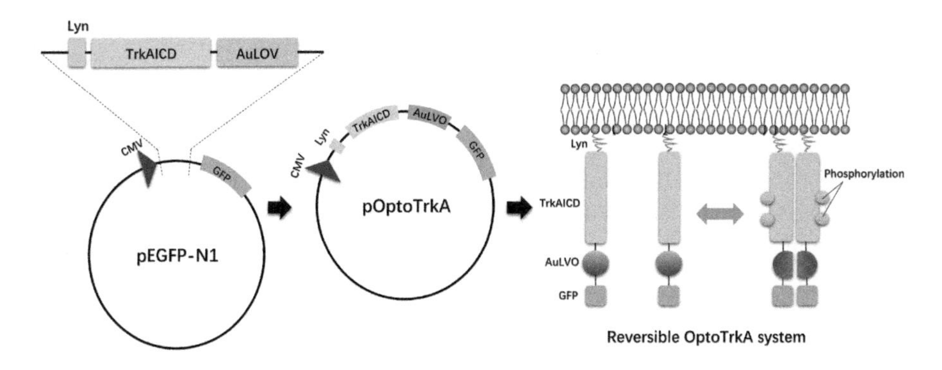

FIGURE 5.7 Schematic illustration showing the construction of OptoTrkA plasmid and the working principle of this optogenetic tool.

5.4.4 Light Stimulation

To stimulate the OptoTrkA activity in the transfected HSCs, a well-established cold light LED illuminator which emitted 450 nm wave blue light at 0.2 mW/cm^2 is placed under the culture plates. The light cycle is set at 10 min on/50 min off, which can sufficiently induce TrkA activity with minimal phototoxicity. However, light intensity and cycle may need to be adjusted when the culture condition is changed with different culture dish or cell type.

5.4.5 Sample Collection and Analysis

The cell migration ability is tested by wound healing assay in which a cell-free gap is created by physical exclusion using a commercialized culture insert from ibidi following the manufacturer's protocol. To avoid interference from cell proliferation, cells are starved in serum-free culture medium.

The effect of light stimulated TrkA on cell proliferation is illustrated by BrdU (5-bromo-2'-deoxyuridine) labeling assay. The 10 μM BrdU reagent is added to the culture medium 1 hour before cell harvest. Then, the cells are fixed with 4% PFA and pre-treated with 2 M HCl at 37°C for 30 min followed by immunostaining using anti-BrdU antibody.

To induce the differentiation of HSCs into neuronal cells, the cells are transferred to a modified neuronal differentiation medium containing DMEM/F12, 1 × B27 supplement, 1 × N2 supplement, 20 ng/mL bone morphogenetic protein 2 (BMP2), and 1 μM all-trans-retinoic acid for 3 days. To induce differentiation of HSCs into glial cells, the cells are transferred to a modified glial differentiation medium containing DMEM/F12, 2 mM L-glutamine, 2 ng/mL insulin, 1 × B27 supplement, 1 × N2 supplement, and 50 ng/mL BMP2 for 3 days. After induction, the cells are fixed by 4% PFA followed by immunostaining using neuronal and glial cell markers, respectively.

The signaling pathways activated by light stimulated TrkA, including Erk1/2 and Akt, are determined with western blotting analysis. After 12 hours light exposure, total proteins of OptoTrkA transfected HSCs are extracted in RIPA buffer with protease inhibitor cocktail. To ascertain whether this light-induced TrkA activity can be switched off, the light-exposed HSC are placed in the dark for another 24 hours followed by western blotting assay (Figure 5.8).

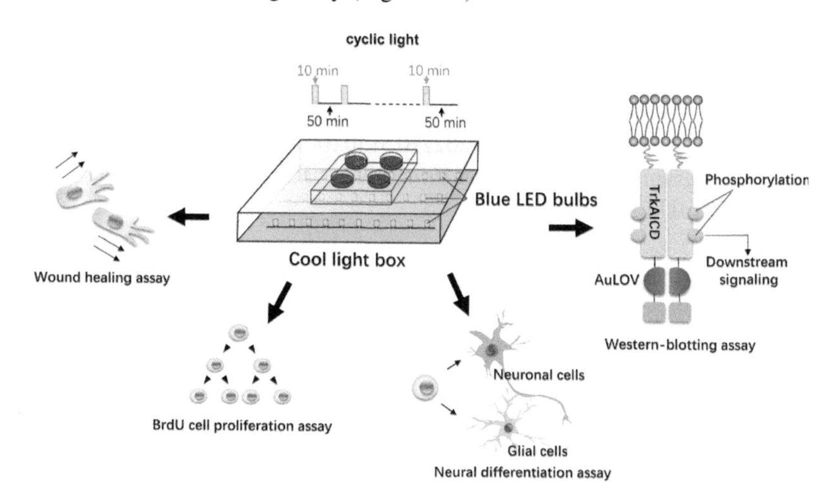

FIGURE 5.8 Schematic workflow showing the molecular and cellular assays performed in determining the function of OptoTrkA in HSCs. With a cooled blue-light illustrator, the TrkA activity is stimulated by cyclic illumination. The migratory ability is determined by a wound healing assay, while the cell proliferation and neural differentiation are elucidated by immunofluorescence staining using BrdU and with specific antibodies targeting neural markers, respectively. The activity of the downstream signaling pathway induced by the OptoTrkA tool is detected by Western blot.

5.4.6 Results and Perspectives

HSCs with light-stimulated TrkA activity show enhanced migratory ability, which can potentially benefit cell grafts colonization in transplant recipient tissue where there is a need to regenerate neurons in large areas or across long distances. Light-induced activation of TrkA promotes the proliferation of HSCs, which could improve cell survival and increase the population size of cell grafts after transplantation. Additionally, OptoTrkA activation significantly accelerates HSC differentiation toward neuronal and glial lineages. Taken together, these results suggest that reversible activation of OptoTrkA in HSCs could lead to therapeutic strategy for neural regeneration in future clinical applications.

5.5 CONCLUDING REMARKS

Dissection of the dynamics of signaling pathways could yield valuable insight into how temporal kinetics of signaling activity is translated to functional outcomes. To date, most pharmacological and genetic tools do not achieve sufficient resolution to precisely control signaling dynamics. Opsin-free optogenetics provides an alternative strategy to address this challenge. As illustrated in this chapter, OptoRaf1 and OptoTrkA have been developed to enable precise *in vitro* manipulation of neural progenitor differentiation and HSC trans-differentiation, respectively. These tools can be further applied *in vivo*, while similar tools could be designed to study the dynamics of other signaling pathways based on the understanding of their activation mechanism.

ACKNOWLEDGMENTS

This work was supported by research grants from the National Natural Science Foundation of China (NSFC 81971309 and 32170980), Guangdong Basic and Applied Basic Research Foundation (2022B1515020012, and Shenzhen Fundamental Research Program (JCYJ20190809161405495, JCYJ20210324123212035 and RCYX20200714114644167).

REFERENCES

1. Santos, S.D., P.J. Verveer, and P.I. Bastiaens, Growth factor-induced MAPK network topology shapes Erk response determining PC-12 cell fate. *Nat Cell Biol*, 2007. **9**(3): pp. 324–330.
2. Deisseroth, K., Optogenetics. *Nat Methods*, 2011. **8**(1): pp. 26–29.
3. Airan, R.D., et al., Temporally precise in vivo control of intracellular signalling. *Nature*, 2009. **458**(7241): pp. 1025–1029.
4. Zhang, K. and B. Cui, Optogenetic control of intracellular signaling pathways. *Trends Biotechnol*, 2015. **33**(2): pp. 92–100.
5. Khamo, J.S., et al., Applications of optobiology in intact cells and multicellular organisms. *J Mol Biol*, 2017. **429**(20): pp. 2999–3017.
6. Namihira, M. and K. Nakashima, Mechanisms of astrocytogenesis in the mammalian brain. *Curr Opin Neurobiol*, 2013. **23**(6): pp. 921–927.

7. Qian, X., et al., Timing of CNS cell generation: a programmed sequence of neuron and glial cell production from isolated murine cortical stem cells. *Neuron*, 2000. **28**(1): pp. 69–80.
8. Lavoie, H. and M. Therrien, Regulation of RAF protein kinases in ERK signalling. *Nat Rev Mol Cell Biol*, 2015. **16**(5): pp. 281–298.
9. Terrell, E.M. and D.K. Morrison, Ras-mediated activation of the raf family kinases. *Cold Spring Harb Perspect Med*, 2019. **9**(1).
10. Tien, A.C., et al., Regulated temporal-spatial astrocyte precursor cell proliferation involves BRAF signalling in mammalian spinal cord. *Development*, 2012. **139**(14): pp. 2477–2487.
11. Urosevic, J., et al., Constitutive activation of B-Raf in the mouse germ line provides a model for human cardio-facio-cutaneous syndrome. *Proc Natl Acad Sci U S A*, 2011. **108**(12): pp. 5015–5020.
12. Holter, M.C., et al., The Noonan Syndrome-linked Raf1L613V mutation drives increased glial number in the mouse cortex and enhanced learning. *PLoS Genet*, 2019. **15**(4): p. e1008108.
13. Kong, S.Y., et al., The histone demethylase KDM5A is required for the repression of astrocytogenesis and regulated by the translational machinery in neural progenitor cells. *Faseb J*, 2018. **32**(2): pp. 1108–1119.
14. Li, X., et al., MEK is a key regulator of gliogenesis in the developing brain. *Neuron*, 2012. **75**(6): pp. 1035–1050.
15. Rhee, Y.H., et al., Neural stem cells secrete factors facilitating brain regeneration upon constitutive Raf-Erk activation. *Sci Rep*, 2016. **6**: p. 32025.
16. Magarinos, M., et al., RAF kinase activity regulates neuroepithelial cell proliferation and neuronal progenitor cell differentiation during early inner ear development. *PLoS One*, 2010. **5**(12): p. e14435.
17. Kolkova, K., et al., Neural cell adhesion molecule-stimulated neurite outgrowth depends on activation of protein kinase C and the Ras-mitogen-activated protein kinase pathway. *J Neurosci*, 2000. **20**(6): pp. 2238–2246.
18. Menard, C., et al., An essential role for a MEK-C/EBP pathway during growth factor-regulated cortical neurogenesis. *Neuron*, 2002. **36**(4): pp. 597–610.
19. Zhang, K., et al., Light-mediated kinetic control reveals the temporal effect of the Raf/MEK/ERK pathway in PC12 cell neurite outgrowth. *PLoS One*, 2014. **9**(3): p. e92917.
20. Su, Y., et al., Early but not delayed optogenetic RAF activation promotes astrocytogenesis in mouse neural progenitors. *J Mol Biol*, 2020. **432**(16): pp. 4358–4368.
21. Hu, K., All roads lead to induced pluripotent stem cells: the technologies of iPSC generation. *Stem Cells Dev*, 2014. **23**(12): pp. 1285–1300.
22. Okita, K., T. Ichisaka, and S. Yamanaka, Generation of germline-competent induced pluripotent stem cells. *Nature*, 2007. **448**(7151): pp. 313–317.
23. Ben-David, U. and N. Benvenisty, The tumorigenicity of human embryonic and induced pluripotent stem cells. *Nat Rev Cancer*, 2011. **11**(4): pp. 268–277.
24. Owczarczyk-Saczonek, A., et al., Therapeutic potential of stem cells in follicle regeneration. *Stem Cells Int*, 2018. **2018**: p. 1049641.
25. Gho, C.G., et al., Isolation, expansion and neural differentiation of stem cells from human plucked hair: A further step towards autologous nerve recovery. *Cytotechnology*, 2016. **68**(5): pp. 1849–1858.
26. Obara, K., et al., Hair-follicle-associated pluripotent stem cells derived from cryopreserved intact human hair follicles sustain multilineage differentiation potential. *Sci Rep*, 2019. **9**(1): p. 9326.
27. Maness, L.M., et al., The neurotrophins and their receptors: structure, function, and neuropathology. *Neurosci Biobehav Rev*, 1994. **18**(1): pp. 143–159.

28. Huang, E.J. and L.F. Reichardt, Neurotrophins: Roles in neuronal development and function. *Annu Rev Neurosci*, 2001. **24**: pp. 677–736.
29. Huang, T., et al., Optogenetically controlled trka activity improves the regenerative capacity of hair-follicle-derived stem cells to differentiate into neurons and glia. *Adv Biol (Weinh)*, 2021. **5**(5): p. e2000134.
30. Zhang, Y., et al., Purification and characterization of progenitor and mature human astrocytes reveals transcriptional and functional differences with mouse. *Neuron*, 2016. **89**(1): pp. 37–53.

6 An Optogenetic Toolbox for Remote Control of Programmed Cell Death

Ningxia Zhang and Ji Jing

CONTENTS

6.1 Introduction 85
6.2 An Optogenetic Toolbox for Precise Control of Cell Death Signaling and Cell Fate 88
6.2.1 Optogenetic Control of Apoptosis 88
6.2.2 Precise Control of Necroptotic Signaling 89
6.2.3 Optogenetic Control of Pyroptotic Cell Death 91
6.2.4 Wireless Optogenetic Control of Cell Death *in vivo* 93
6.3 Conclusions and Future Directions or Perspectives 94
Acknowledgments 94
References 95

6.1 INTRODUCTION

Optogenetics is defined as a means of using genetically encoded light-sensitive proteins to noninvasively manipulate cells with light.[1,2] Channelrhodopsin-2 (ChR2) was first applied to neurons to control action potentials. This kind of microbial opsin-based optogenetic tool has been widely used in neuroscience studies.[3–5] With multiple non-opsin-based light-sensitive proteins from bacteria and plants being discovered (Figure 6.1, Table 6.1),[6–8] optogenetics has expanded beyond neuroscience to regulate the physiological activities of non-neuronal cells, such as controlling ion signaling pathways (calcium, potassium ions),[9–11] light-mediated gene editing and expression,[8] engineering for immune cell therapies,[12,13] directional regulation of the "behavior" of cells and animals,[14–16] and photo-controlling cell fate (Table 6.2).[14,17,18]

Light-oxygen-voltage (LOV) domains[19] and cryptochrome[20] are among the most widely used non-opsin blue light-responsive photoreceptors to control intracellular signaling. The LOV domain contains a conserved Per-Arnt-Sim (PAS) core domain bound with flavin mononucleotide (FMN) and a C-terminal Jα-helix. The cofactor FMN has an absorption maximum around 450 nm and can induce an intramolecular conformational change under blue light.[21] A protein of interest (POI) can be fused to the Jα-helix at the C-terminus, thereby enabling the caging of POI in the dark and release of POI upon blue light irradiation. To allow caging of proteins requiring

DOI: 10.1201/b22823-6

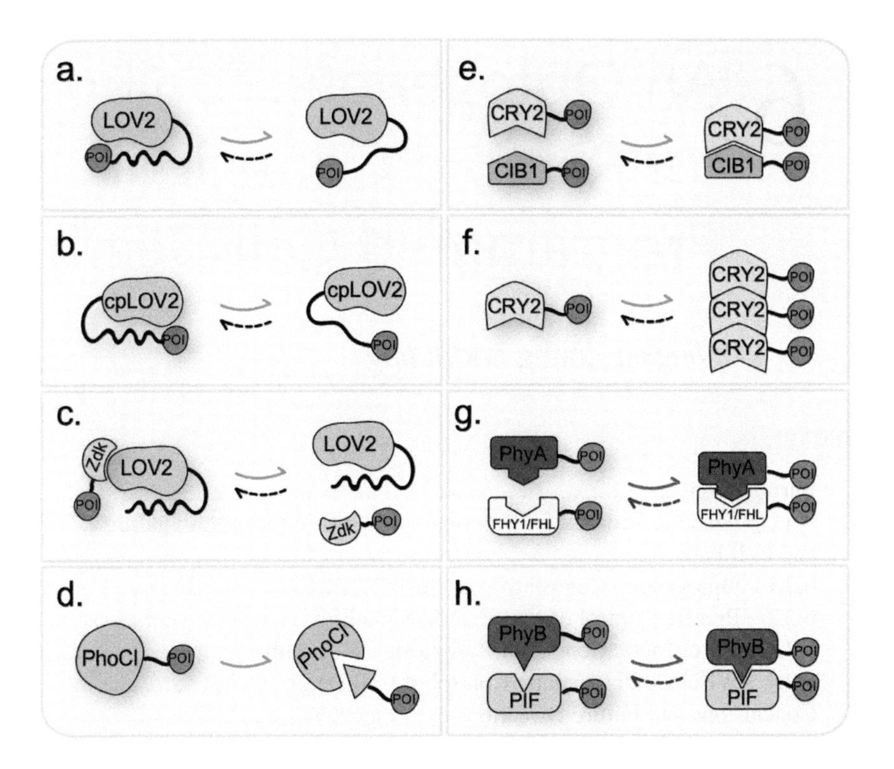

FIGURE 6.1 Schematic representation of opsin-free photoreceptors. (a) The Jα-helix of LOV2 adopts a well-folded helical structure in the dark and becomes disordered upon blue light irradiation, thereby uncaging the protein of interest (POI) fused to the C-terminal end. (b) A circularly permuted LOV2 variant with its Jα-helix located at the N-terminus. (c) In the LOVTRAP system, Zdk tightly binds to LOV2 in the dark and undocks from LOV2 upon blue light irradiation. (d) PhoCl undergoes violet light-inducible self-cleavage to produce two fragments. (e) CRY2 and its interaction partner CIB1 form heterodimer under blue light irradiation and dissociate in the dark. (f) CRY2 undergoes homo-oligomerization upon blue light irradiation. (g) PhyA and its interaction partner FHY1/FHL form heterodimer under red light irradiation and dissociate under far-red light. (h) PhyB and its interaction partner PIF form heterodimers under red light irradiation and dissociates under far-red light.

TABLE 6.1

Overview of Representative Non-Opsin Photoreceptors

Photoreceptor	Trigger	Working mechanism	Refs.
LOV2	Blue light	Light-inducible unfolding	19, 21
cpLOV2	Blue light	Light-inducible unfolding	14, 22
LOVTRAP	Blue light	Light-induced dissociation	49
PhoCI	Violet	Photo-triggered self-cleavage	48
CRY2-CIB1	Blue light	Light-dependent heterodimerization	7
CRY2	Blue light	Light-triggered self-oligomerization	20
PhyA-FHY1/FHL	Red light	Light-dependent heterodimerization	56
PhyB-PIF	Red light	Light-dependent heterodimerization	57

TABLE 6.2
Exemplary Optogenetic Tools for Programmed Cell Death Manipulation

Cell death	Photoreceptor	Trigger	Targets	Domain features	Refs.
Apoptosis	CRY2-PHR	Blue	BAX^{S184E}	S184E	58
	CRY2	Blue	Caspase-8	ΔCARD (92–416)	18
	CRY2	Blue	Caspase-9	ΔDED (177–479)	18
Necroptosis	CRY2	Blue	RIPK1	Full length	17
	CRY2	Blue	RIPK3	Wildtype	17
				ΔRHIM	18
	CRY2	Blue	MLKL-NT	(1–125)-3A	17
	cpLOV2	Blue	MLKL-NT	(1–178)	14
Pyroptosis	CRY2	Blue	Caspase-1	ΔCARD (92–404)	18
	CRY2	Blue	Caspase-4	ΔCARD (92–377)	18
	CRY2	Blue	Caspase-5	ΔCARD (90–435)	18
	LOVTRAP	Blue	mGSDMD	Wildtype	17
	PhoCl	Violet	hGSDMD	Wildtype	48
	cpLOV2	Blue	mGSDMD-NT	(1–276)-4A	17

free N-termini for optimal functionality, circularly permuted LOV2 (cpLOV2) was generated for more versatile optogenetic engineering.[14,22]

The cryptochrome protein Cryptochrome Circadian Regulator 2 (CRY2) from *Arabidopsis thaliana* undergoes conformational changes under blue light via flavin adenine dinucleotide (FAD) photoreduction, resulting in the formation of CRY2-CIB1 heterodimers or CRY2 homo-oligomers in the presence or absence of cryptochrome-interacting basic-helix-loop-helix (CIB1).[7] This unique property makes it possible to achieve optical manipulation of protein-protein interactions or caging/uncaging events, thereby enabling precise control of signal transduction pathways and cell fate in a spatiotemporal dependent way without external ligands.

With the discovery of these non-opsin-based light-sensitive proteins, a growing number of optogenetic tools have been developed to manipulate physiological processes, among which the cell death pathways are of enormous interest. Some cells choose to commit "suicide" actively in a carefully controlled way, which is known as programmed cell death (PCD), thus maintaining a delicate balance between cell survival and cell death.[23] PCD is characterized by precise signaling cascades executed by a diverse array of interconnected proteins.[24] This reliance on specialized intracellular signaling molecular machinery implies that any subtle changes in the involved pathways might lead to uncontrolled cell death, which subsequently contributes to diseases ranging from autoimmune disorders to cancer. Conversely, it also means that PCD can be manipulated by optogenetics with high spatiotemporal precision to intervene in diseases associated with dysregulated PCD. In this chapter, we will focus mainly on the applications of non-opsin optogenetics in light-induced programmed cell death, including apoptosis, necroptosis, and pyroptosis.

6.2 AN OPTOGENETIC TOOLBOX FOR PRECISE CONTROL OF CELL DEATH SIGNALING AND CELL FATE

As the latest Nomenclature Committee on Cell Death (NCCD) guideline shows, cell death can be divided into accidental cell death (ACD) and PCD according to signal dependency. According to the morphological characteristics and the underlying molecular mechanisms, PCD can be further divided into different types of cell death, including apoptosis, necroptosis, ferroptosis, pyroptosis, parthanatos, entotic cell death, NETotic cell death, lysosome-dependent cell death, autophagy-dependent cell death, and immunogenic cell death.[23]

6.2.1 Optogenetic Control of Apoptosis

Apoptosis is the earliest proposed PCD, which is marked by cell shrinkage, membrane blistering, positional organelles loss, DNA concentration, and apoptotic body formation.[25] Two main pathways of apoptosis, intrinsic apoptosis and extrinsic apoptosis, have been identified.[26] The most indispensable step of intrinsic apoptosis is the release of mitochondrial proteins (such as cytochrome c) caused by mitochondrial outer membrane permeabilization (MOMP).[27] This process is strictly regulated by the BCL2 family of proteins, including pro-apoptotic members (such as BAX, BAK, BID, BIM, and PUMA) and anti-apoptotic members (such as BCL2, BCL-XL, and BCL-W), among which oligomerization of BAX and BAK on the outer mitochondrial membrane (OMM) poses a direct effect on membrane permeabilization.[23,28] Both BAX and BAK can independently promote apoptosis.[29–31] Compared to OMM-resident BAK, BAX is mainly located in the cytosol in an inactive conformation and translocates toward mitochondria with the help of other pro-apoptotic proteins.[29,32] Normally, this MOMP-dependent death is strictly and precisely regulated according to the different physiological states of cells.[31] Hence, if the oligomerization of BAX to OMM can be manipulated artificially, researchers can initiate the execution of PCD.

To enable light-inducible recruitment of pro-apoptoic proteins toward OMM, an optical dimerizer composed of CRY2 and CIB1 was used to craft an optogenetic suicide device. The device consists of two key components: (i) Tom20-CIB1-GFP is used to anchor the CIB1 to the surface of the OMM via Tom20; (ii) CRY2-mCh-BAX^{S184E}, which remains evenly distributed in the cytosol in the dark but translocates toward OMM via CRY2-CIB1 heterodimerization under blue light illumination. The light-inducible recruitment of BAX^{S184E} toward OMM leads to MOMP and triggers the downstream caspase-9 oligomerization to activate the executioner caspases (mainly caspase-3) (Figure 6.2a).[33] In addition, extrinsic apoptosis is executed through the caspase-3 protein after receiving a signal detected by plasma membrane receptors, and the difference is that caspase-8 serves as the transmitter in this extrinsic pathway.[23] Utilizing this feature, another optogenetic tool, optoCaspase-8/9, has been developed to regulate apoptosis, which is built upon CRY2-induced polymerization after exposure to blue light (Figure 6.2b-c).[18] The caspase recruitment domain (CARD) of caspase-8 and the death effector domain (DED) of caspase-9 are involved in protein-protein interactions in inflammation and

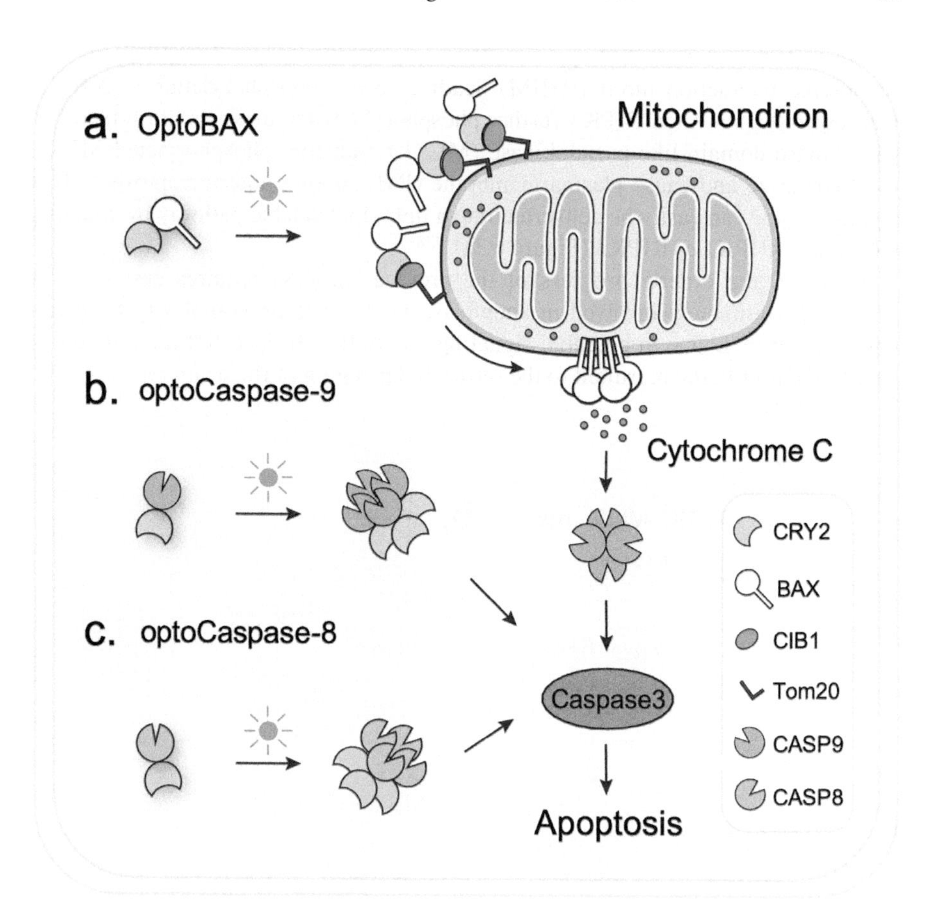

FIGURE 6.2 Optogenetic approaches to control apoptosis. (a) CIB1 is anchored to the outer mitochondrial membrane (OMM) via fusion to Tom20 while a BAX mutant is fused to CRY2 in the cytosol. Upon blue light illumination, CRY2-BAX undergoes oligomerization and further engages CIB1 in the OMM to promote apoptosis. (b-c) Caspase-8/9 is fused with CRY2 to form oligomers after blue light illumination, thereby mimicking caspase-8/9 induced activation of caspase-3. CASP8, caspase-8; CASP9, caspase-9.

apoptosis.[34,35] Interestingly, the apoptotic efficiency is improved by using CARD/DED-deficient optoCaspase-8/9 compared with full-length optoCaspase-8/9, whose apoptotic potential could be reduced via interaction with other proteins.

6.2.2 Precise Control of Necroptotic Signaling

Necrosis has been generally regarded as merely accidental cell death.[36] However, despite showing the morphological characteristics reminiscent of necrosis, necroptosis is strictly regulated by programs that can be simply summarized as the RIPK1-RIPK3-MLKL cascade pathway.[36] Upon death receptor activation, receptor-interacting serine/threonine-protein kinase 1 (RIPK1) undergoes self-oligomerization and recruits receptor-interacting serine/threonine-protein kinase 3 (RIPK3) via RIP homotypic interaction motif (RHIM), leading to autophosphorylation of RIPK3. Consequently, activated RIPK3 further phosphorylates the downstream mixed lineage kinase domain-like pseudokinase (MLKL), inducing phosphorylated MLKL to oligomerize and cause plasma membrane (PM) rupture and necroptosis.[37] This signal cascade can easily be converted into a light-controllable pathway by coupling CRY2 with RIPK1 or RIPK3 (Figure 6.3a).[17,18]

However, the orderly progress of the previous process requires caspase-8 as an enzyme intimately involved in apoptosis, thereby adding complexity to different forms of PCD due to potential signaling crosstalk.[38] In fact, RIPK1 is ubiquitinated first after being recruited to the intracellular region of the death receptor. The

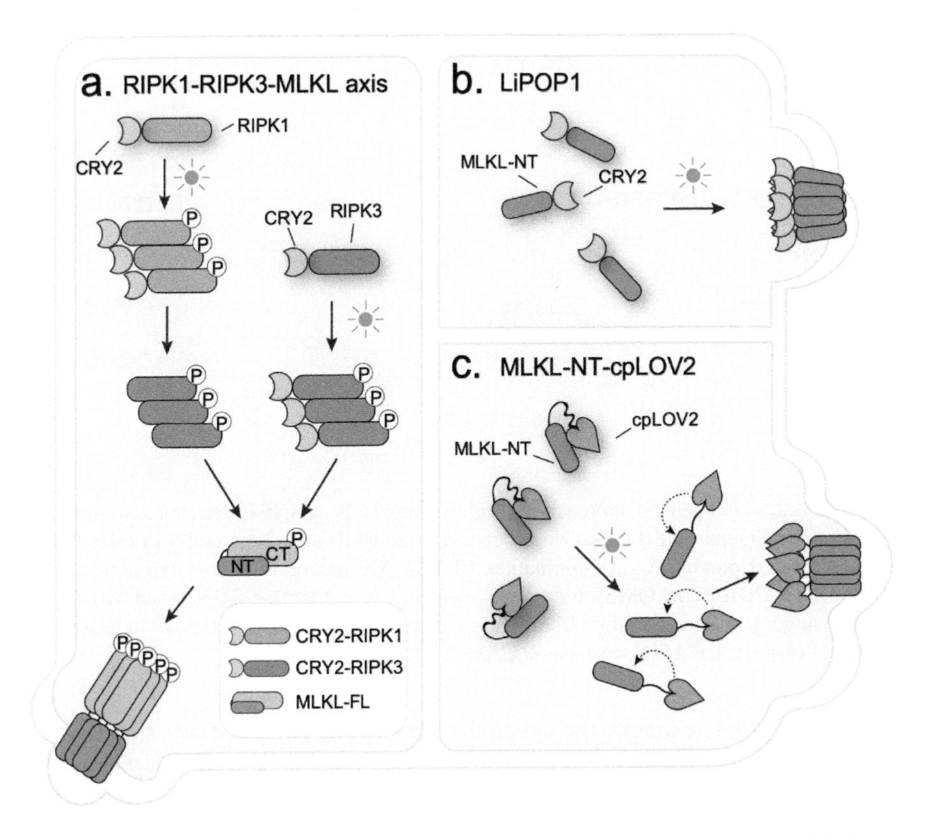

FIGURE 6.3 Optogenetic approaches to control necroptosis. (a) RIPK1 or RIPK3 is fused with CRY2 to form oligomers under blue light, inducing phosphorylated MLKL to undergo oligomerization and cause the plasma membrane rupture and necroptosis. (b) LiPOP1 design. An optimized N-terminal domain of MLKL (MLKL-NT) is fused to the N-terminus of CRY2 to enable light-inducible MLKL-NT oligomerization to disrupt plasma membrane directly. (c) The cell-killing N-terminal fragment of MLKL is released under blue light from the MLKL-NT-cpLOV2 chimera protein. P, phosphorylation site; NT, the N-terminal domain of MLKL; CT, the C-terminal domain of MLKL; FL, full length.

ubiquitinated RIPK1 participates in the downstream complex assembly to activate the NF-kB pathway and promote cell survival. Further, the RIPK1-FADD-caspase-8 complex is formed to activate apoptosis. In this way, RIPK1 acts as a repressor rather than an activator of RIPK3. Deubiquitinases remove the ubiquitin group of RIPK1, thus inducing RIPK1 assembling and recruitment of RIPK3.[39]

Therefore, a better way to induce necroptosis is via direct photo-control of MLKL self-oligomerization by CRY2. The MLKL N-terminal region (MLKL-NT), particularly the four helical bundle domain (4HBD), plays a decisive role in causing PM rupture.[40] After a careful screening via mutagenesis, the best construct, MLKL (1–125 with mutations H15A/K16A/R17A), with high photo-responsiveness and least basal activity, was selected and named as light-induced non-apoptotic tool 1 (LiPOP1, Figure 6.3b).[17] Under pulsed blue light illumination, LiPOP1 causes cell death within 30 minutes in mammalian cells. Since the wildtype MLKL-NT (1–125) has considerable cytotoxicity even without light stimulation, LOV2 was used as a fusion tag to limit its activity in the dark.[17] However, appended MLKL-NT at the C terminus of LOV2 (LOV2-MLKL-NT) seemed to negatively affect the cell-killing capability,[14] likely because a free N-terminus of MLKL-NT is required for efficient PM rupture. To overcome this barrier, cpLOV2 allows the fusion of MLKL-NT with its N-terminus unmodified while still caging MLKL-NT was used (Figure 6.3c).[14] Indeed, by playing with the relative positions among the PAS core domain, the Jα helix of LOV2, and the effector domain, cpLOV2 permits the fusion of an effector domain upstream of the photosensory module without affecting the effector activity upon light stimulation.

6.2.3 Optogenetic Control of Pyroptotic Cell Death

Pyroptosis is an inflammatory form of programmed cell death that also requires caspase activation and subsequent PM pores formation induced by the gasdermin family proteins.[41] The gasdermin protein family contains five members, gasdermin A to E, which share highly conserved N-terminal and C-terminal sequences separated by a variable linker.[42] Among these, GSDMD is most well studied, which acts as the downstream effector molecule of pyroptosis.[43] Accumulating evidence suggests that GSDMD exists in an inactive state because the autoinhibition of its C-terminal repressor domain folds back on its N-terminal pore-forming domain. The activated caspases cleave the linker region of GSDMD and release the pore-forming N-terminal domain, which undergoes oligomerization and translocation to the PM to perforate PM, thus leading to cell death.[44] Caspases involved in this process include caspase-1, caspase-3, human caspase-4, and caspase-5 (corresponding to murine caspase-11), among which caspases-4/5/11 activation could be triggered by directly binding of cytosolic bacterial lipopolysaccharides (LPS) while caspase-1 by inflammasome, both activated via oligomerization.[45] In addition, caspase-1 also processes the inactive pro-forms of inflammatory cytokines such as interleukin-1β (IL-1β) and interleukin-18 (IL-18) into active forms, which are released from the GSDMD pores to initiate the downstream inflammatory cascade.[23] In the absence of GSDMD, caspase-1 initiates apoptosis via the Bid-caspase 9-caspase 3 axis.[46] Conversely, apoptosis can also be switched to pyroptosis under certain circumstances, such as GSDME cleavage due to caspase-3 activation.[47]

Similar to optoCaspase-8/9 (Figure 6.2b-c) that are designed for light-controlled apoptosis, CRY2 can be introduced into caspase-1/4/5 to form light-controlled pyroptotic system (Figure 6.4a).[18] However, due to the crosstalk between multiple caspases, this strategy inevitably fails to discriminate between apoptosis and pyroptosis. Hence, optogenetic tools were designed to directly control the liberation of

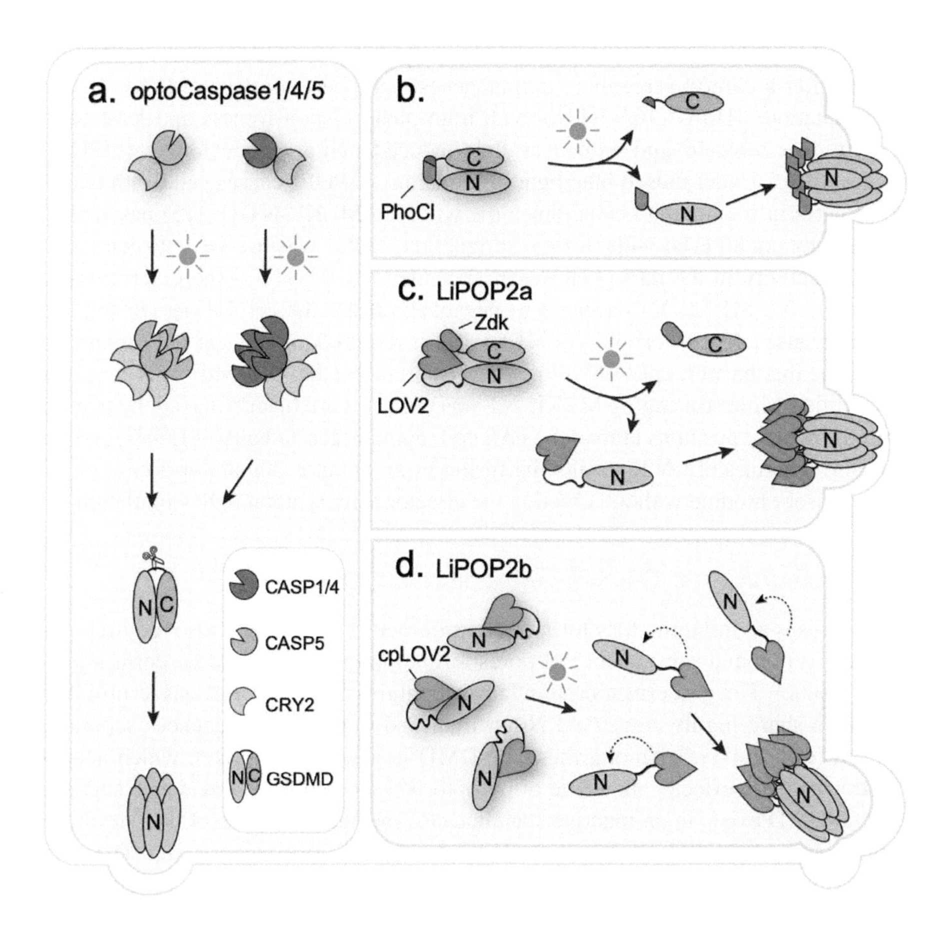

FIGURE 6.4 Optogenetic approaches to control Pyroptosis. (a) Caspase-1/4/5 is fused with CRY2 to initiate cleavage of GSDMD in response to blue light, thereby releasing the pore-forming GSDMD N-terminal domain (GSDMD-NT) to induce pyroptosis. (b) PhoCl is inserted between GSDMD-NT and GSDMD-CT, which enable light-inducible self-cleavage to produce two fragments after blue light exposure, liberating the GSDMD-NT for membrane perforation. (c-d) The LiPOP2 system. In LiPOP2a, the pore-forming activity of GSDMD-NT is inhibited by GSDMD-CT via the LOV2-Zdk association in the dark. Zdk dissociates from LOV, thereby exposing GSDMD-NT to restore its pore-forming capability upon blue light illumination. For LiPOP2b, an optimized GSDMD-NT mutant is fused with cpLOV2 and remains caged in the dark. Under blue light stimulation, the liberated GSDMD-NT oligomerizes to ultimately cause pyroptosis. CASP1/4, caspase-1/4; CASP5, caspase-5; N, the N-terminal domain of MLKL; C, the C-terminal domain of MLKL.

the N-terminal domain of GSDMD. Three different engineering approaches were adopted. Diffusible optogenetic GSDMD (PhoDer) was developed to study the dynamics of gasdermin pores. This tool inserted a light-cleavable protein PhoCl between GSDMD-NT and GSDMD-CT, which underwent cleavage with the subsequent irreversible release of the N-terminal pore-forming domain upon light irradiation (Figure 6.4b).[48] The second strategy involves light-inducible murine GSDMD (mGSDMD) autoinhibition mimicry, is achieved by fusing GSDMD-NT and GSDMD-CT to LOV2 Trap and Release of Protein (LOVTRAP), a light-inducible dissociation system. Z-dark (Zdk) is a small protein bearing a three-helix bundle, which docks to LOV2 in the dark and dissociates from LOV2 upon light stimulation.[49] By fusing the mGSDMD-NT and mGSDMD-CT with LOV2 and Zdk respectively, one could mimic the autoinhibition of mGSDMD-CT on mGSDMD-NT in the dark. Light-induced LOV2-Zdk dissociation disrupts the interaction between GSDMD-NT and GSDMD-CT, thereby releasing the N-terminus to perform pore-forming function (Figure 6.4c).[17]

The third approach involves the conditional caging of mGSDMD-NT. Considering that the protein tag fusion at the N-terminus of mGSDMD tends to impair its pore-forming function, mGSDMD-NT was appended in front of cpLOV2 instead of after the C-terminus of LOV2. The active site of mGSDMD-NT was masked by cpLOV2 before light illumination, which mimics the inhibitory role of GSDMD-CT to a certain extent. To reduce the toxicity of GSDMD-NT in the dark state, several K/R-to-A mutations were introduced to neutralize the positive charges in the postulated pore-forming region of mGSDMD-NT, with the optimized 4A mutant (R138A/K146A/R152A/R154A) showing low cytotoxicity in the dark but regaining potent pore-forming capacity in the lit state (Figure 6.4d).[17]

6.2.4 Wireless Optogenetic Control of Cell Death *in vivo*

Individually, the death of a cell means the end of life. However, for a living organism, the death of one cell or one cell population might benefit the survival of the remaining cells. PCD is also considered a self-protection mechanism of cells in response to different stresses, participating in various physiological and pathological regulatory processes and maintaining tissue homeostasis. Disruption of the intricate balance between cell survival and cell death can lead to uncontrolled cell growth and malignant transformation. Optogenetics offers a new avenue to remotely control cell death with exquisite temporal and spatial precision, and therefore holds great promise in cancer treatment. Nonetheless, two key issues need to be addressed before blue photoreceptor-based optogenetics can be translated into the clinical setting: (i) the shallow tissue penetration ability of blue light and (ii) the inconvenience of light delivery to regions of interest in a living organism.

Limited by the short wavelength of blue light, which can only achieve a limited penetration of up to 0.5–1 mm in the human skin,[17,50] it remains challenging to use non-opsin optogenetic tools like LiPOP1 to achieve optical control of cell death *in vivo*. The combination of LiPOP1 with upconversion nanoparticles (UCNPs) provides a tentative solution to this problem.[9] UCNPs act as nano-transducers that can convert the deep tissue-penetrable near-infrared (NIR) light into blue light emission, so as to activate LiPOP1 *in vivo*. In this way, wireless light-controlled suicide of subcutaneous xenograft tumors expressing LiPOP1 can be achieved.[17] For deeply buried tissues, NanoLuc-mediated luminescence-aided optogenetic stimulation (NanoLOGS) platform offers a more effective strategy to activate LiPOP1 *in vivo.* LiPOP1-NanoLuc relies on the substrate fluorofurimazine (FFz) to produce bright luminescence peaked at the wavelength of 450–470 nm, which perfectly matches the photoactivation window of LOV2 or CRY2-based optogenetic tools.[17,51] However, this strategy partially sacrifices the temporal precision of optogenetic tools and requires further optimization to avoid potential side effects arising from the chemical substrate furimazine or its derivatives.

6.3 CONCLUSIONS AND FUTURE DIRECTIONS OR PERSPECTIVES

Optogenetics provides an efficient way to study and exploit programmed cell death, which is often complicated by signaling crosstalk among different pathways. Opsin-free optogenetic control of PCD signaling provides a novel method to dissect the signaling circuits of different types of cell death. Worthy to note, pyroptotic cells tend to release pro-inflammatory factors to affect neighboring cells and recruit immune cells, thereby creating a "hot" tumor microenvironment (TME) to facilitate tumor killing. Only a small fraction of pyroptotic tumor cells is required to trigger a self-amplifying inflammatory response, thus improving the tumor microenvironment and activating robust cytotoxic T cell-mediated anti-tumor immune response.[52] Ideally, light can be exploited as the trigger to initiate localized pyroptosis, thereby instructing more efficient immune cell infiltration into TME to turn "cold" tumors into "hot" ones.[17]

Optogenetic tools described herein could also be exploited as suicide-inducing devices to develop intelligent cell-based therapy.[53] CAR T-cell therapy emerges as a highly promising approach for cancer treatment, which combines the antigen-binding properties of antibodies with the anti-tumor effector function of T cells, using engineered T cells to clear tumor cells.[54] However, uncontrolled CAR T-cell activation can lead to adverse effects, such as cytokine release syndrome.[55] Modifying the existing CAR T-cells with a light-inducible suicide gene could provide a safety switch to enable the conditional removal of over-activated therapeutic immune cells.[17] It is anticipated that LiPOP-like tools will find more applications in optogenetic immunomodulation and the design of smart cellular therapies in the forthcoming years.

ACKNOWLEDGMENTS

We thank the financial supports from the Key Laboratory of Prevention, Diagnosis and Therapy of Upper Gastrointestinal Cancer of Zhejiang Province (2022E10021), and the start-up fund from The Cancer Hospital of the University of Chinese Academy of Sciences (Zhejiang Cancer Hospital) and Institute of Basic Medicine and Cancer (to JJ).

REFERENCES

1. Mansouri, M.; Strittmatter, T.; Fussenegger, M., Light-controlled mammalian cells and their therapeutic applications in synthetic biology. *Adv Sci (Weinh)* **2019**, *6* (1), 1800952.
2. News, S., Insights of the decade. Stepping away from the trees for a look at the forest. Introduction. *Science* **2010**, *330* (6011), 1612–1613.
3. Boyden, E. S.; Zhang, F.; Bamberg, E.; Nagel, G.; Deisseroth, K., Millisecond-timescale, genetically targeted optical control of neural activity. *Nat Neurosci* **2005**, *8* (9), 1263–1268.
4. Tischer, D.; Weiner, O. D., Illuminating cell signalling with optogenetic tools. *Nat Rev Mol Cell Biol* **2014**, *15* (8), 551–558.
5. Deisseroth, K., Optogenetics: 10 years of microbial opsins in neuroscience. *Nat Neurosci* **2015**, *18* (9), 1213–1225.
6. Strickland, D.; Lin, Y.; Wagner, E.; Hope, C. M.; Zayner, J.; Antoniou, C.; Sosnick, T. R.; Weiss, E. L.; Glotzer, M., TULIPs: Tunable, light-controlled interacting protein tags for cell biology. *Nat Methods* **2012**, *9* (4), 379–384.
7. Kennedy, M. J.; Hughes, R. M.; Peteya, L. A.; Schwartz, J. W.; Ehlers, M. D.; Tucker, C. L., Rapid blue-light-mediated induction of protein interactions in living cells. *Nat Methods* **2010**, *7* (12), 973–975.
8. Zhou, Y.; Kong, D.; Wang, X.; Yu, G.; Wu, X.; Guan, N.; Weber, W.; Ye, H., A small and highly sensitive red/far-red optogenetic switch for applications in mammals. *Nat Biotechnol* **2022**, *40* (2), 262–272.
9. He, L.; Zhang, Y.; Ma, G.; Tan, P.; Li, Z.; Zang, S.; Wu, X.; Jing, J.; Fang, S.; Zhou, L.; Wang, Y.; Huang, Y.; Hogan, P. G.; Han, G.; Zhou, Y., Near-infrared photoactivatable control of Ca(2+) signaling and optogenetic immunomodulation. *Elife* **2015**, *4*.
10. Ma, G.; He, L.; Liu, S.; Xie, J.; Huang, Z.; Jing, J.; Lee, Y. T.; Wang, R.; Luo, H.; Han, W.; Huang, Y.; Zhou, Y., Optogenetic engineering to probe the molecular choreography of STIM1-mediated cell signaling. *Nat Commun* **2020**, *11* (1), 1039.
11. Kim, S.; Kyung, T.; Chung, J. H.; Kim, N.; Keum, S.; Lee, J.; Park, H.; Kim, H. M.; Lee, S.; Shin, H. S.; Heo, W. D., Non-invasive optical control of endogenous Ca(2+) channels in awake mice. *Nat Commun* **2020**, *11* (1), 210.
12. Nguyen, N. T.; Huang, K.; Zeng, H.; Jing, J.; Wang, R.; Fang, S.; Chen, J.; Liu, X.; Huang, Z.; You, M. J.; Rao, A.; Huang, Y.; Han, G.; Zhou, Y., Nano-optogenetic engineering of CAR T cells for precision immunotherapy with enhanced safety. *Nat Nanotechnol* **2021**, *16* (12), 1424–1434.
13. Huang, Z.; Wu, Y.; Allen, M. E.; Pan, Y.; Kyriakakis, P.; Lu, S.; Chang, Y. J.; Wang, X.; Chien, S.; Wang, Y., Engineering light-controllable CAR T cells for cancer immunotherapy. *Sci Adv* **2020**, *6* (8), eaay9209.
14. He, L.; Tan, P.; Zhu, L.; Huang, K.; Nguyen, N. T.; Wang, R.; Guo, L.; Li, L.; Yang, Y.; Huang, Z.; Huang, Y.; Han, G.; Wang, J.; Zhou, Y., Circularly permuted LOV2 as a modular photoswitch for optogenetic engineering. *Nat Chem Biol* **2021**, *17* (8), 915–923.
15. Guntas, G.; Hallett, R. A.; Zimmerman, S. P.; Williams, T.; Yumerefendi, H.; Bear, J. E.; Kuhlman, B., Engineering an improved light-induced dimer (iLID) for controlling the localization and activity of signaling proteins. *Proc Natl Acad Sci U S A* **2015**, *112* (1), 112–117.
16. Gradinaru, V.; Thompson, K. R.; Zhang, F.; Mogri, M.; Kay, K.; Schneider, M. B.; Deisseroth, K., Targeting and readout strategies for fast optical neural control in vitro and in vivo. *J Neurosci* **2007**, *27* (52), 14231–14238.
17. He, L.; Huang, Z.; Huang, K.; Chen, R.; Nguyen, N. T.; Wang, R.; Cai, X.; Huang, Z.; Siwko, S.; Walker, J. R.; Han, G.; Zhou, Y.; Jing, J., Optogenetic control of non-apoptotic cell death. *Adv Sci (Weinh)* **2021**, *8* (13), 2100424.

18. Shkarina, K.; Hasel de Carvalho, E.; Santos, J. C.; Leptin, M.; Broz, P., Optogenetic activators of apoptosis, necroptosis and pyroptosis for probing cell death dynamics and bystander cell responses. *bioRxiv* **2021**, 2021.08.31.458313.
19. Lee, J.; Natarajan, M.; Nashine, V. C.; Socolich, M.; Vo, T.; Russ, W. P.; Benkovic, S. J.; Ranganathan, R., Surface sites for engineering allosteric control in proteins. *Science* **2008**, *322* (5900), 438–442.
20. Bugaj, L. J.; Choksi, A. T.; Mesuda, C. K.; Kane, R. S.; Schaffer, D. V., Optogenetic protein clustering and signaling activation in mammalian cells. *Nat Methods* **2013**, *10* (3), 249–252.
21. Wu, Y. I.; Frey, D.; Lungu, O. I.; Jaehrig, A.; Schlichting, I.; Kuhlman, B.; Hahn, K. M., A genetically encoded photoactivatable Rac controls the motility of living cells. *Nature* **2009**, *461* (7260), 104–108.
22. Geng, L.; Shen, J.; Wang, W., Circularly permuted AsLOV2 as an optogenetic module for engineering photoswitchable peptides. *Chem Commun (Camb)* **2021**, *57* (65), 8051–8054.
23. Galluzzi, L.; Vitale, I.; Aaronson, S. A.; Abrams, J. M.; Adam, D.; Agostinis, P.; Alnemri, E. S.; Altucci, L.; Amelio, I.; Andrews, D. W.; Annicchiarico-Petruzzelli, M.; Antonov, A. V.; Arama, E.; Baehrecke, E. H.; Barlev, N. A.; Bazan, N. G.; Bernassola, F.; Bertrand, M. J. M.; Bianchi, K.; Blagosklonny, M. V.; Blomgren, K.; Borner, C.; Boya, P.; Brenner, C.; Campanella, M.; Candi, E.; Carmona-Gutierrez, D.; Cecconi, F.; Chan, F. K.; Chandel, N. S.; Cheng, E. H.; Chipuk, J. E.; Cidlowski, J. A.; Ciechanover, A.; Cohen, G. M.; Conrad, M.; Cubillos-Ruiz, J. R.; Czabotar, P. E.; D'Angiolella, V.; Dawson, T. M.; Dawson, V. L.; De Laurenzi, V.; De Maria, R.; Debatin, K. M.; DeBerardinis, R. J.; Deshmukh, M.; Di Daniele, N.; Di Virgilio, F.; Dixit, V. M.; Dixon, S. J.; Duckett, C. S.; Dynlacht, B. D.; El-Deiry, W. S.; Elrod, J. W.; Fimia, G. M.; Fulda, S.; García-Sáez, A. J.; Garg, A. D.; Garrido, C.; Gavathiotis, E.; Golstein, P.; Gottlieb, E.; Green, D. R.; Greene, L. A.; Gronemeyer, H.; Gross, A.; Hajnoczky, G.; Hardwick, J. M.; Harris, I. S.; Hengartner, M. O.; Hetz, C.; Ichijo, H.; Jäättelä, M.; Joseph, B.; Jost, P. J.; Juin, P. P.; Kaiser, W. J.; Karin, M.; Kaufmann, T.; Kepp, O.; Kimchi, A.; Kitsis, R. N.; Klionsky, D. J.; Knight, R. A.; Kumar, S.; Lee, S. W.; Lemasters, J. J.; Levine, B.; Linkermann, A.; Lipton, S. A.; Lockshin, R. A.; López-Otín, C.; Lowe, S. W.; Luedde, T.; Lugli, E.; MacFarlane, M.; Madeo, F.; Malewicz, M.; Malorni, W.; Manic, G.; Marine, J. C.; Martin, S. J.; Martinou, J. C.; Medema, J. P.; Mehlen, P.; Meier, P.; Melino, S.; Miao, E. A.; Molkentin, J. D.; Moll, U. M.; Muñoz-Pinedo, C.; Nagata, S.; Nuñez, G.; Oberst, A.; Oren, M.; Overholtzer, M.; Pagano, M.; Panaretakis, T.; Pasparakis, M.; Penninger, J. M.; Pereira, D. M.; Pervaiz, S.; Peter, M. E.; Piacentini, M.; Pinton, P.; Prehn, J. H. M.; Puthalakath, H.; Rabinovich, G. A.; Rehm, M.; Rizzuto, R.; Rodrigues, C. M. P.; Rubinsztein, D. C.; Rudel, T.; Ryan, K. M.; Sayan, E.; Scorrano, L.; Shao, F.; Shi, Y.; Silke, J.; Simon, H. U.; Sistigu, A.; Stockwell, B. R.; Strasser, A.; Szabadkai, G.; Tait, S. W. G.; Tang, D.; Tavernarakis, N.; Thorburn, A.; Tsujimoto, Y.; Turk, B.; Vanden Berghe, T.; Vandenabeele, P.; Vander Heiden, M. G.; Villunger, A.; Virgin, H. W.; Vousden, K. H.; Vucic, D.; Wagner, E. F.; Walczak, H.; Wallach, D.; Wang, Y.; Wells, J. A.; Wood, W.; Yuan, J.; Zakeri, Z.; Zhivotovsky, B.; Zitvogel, L.; Melino, G.; Kroemer, G., Molecular mechanisms of cell death: Recommendations of the nomenclature committee on cell death 2018. *Cell Death Differ* **2018**, *25* (3), 486–541.
24. Elmore, S., Apoptosis: A review of programmed cell death. *Toxicol Pathol* **2007**, *35* (4), 495–516.
25. Galluzzi, L.; Maiuri, M. C.; Vitale, I.; Zischka, H.; Castedo, M.; Zitvogel, L.; Kroemer, G., Cell death modalities: Classification and pathophysiological implications. *Cell Death Differ* **2007**, *14* (7), 1237–1243.

26. Schulze-Osthoff, K.; Ferrari, D.; Los, M.; Wesselborg, S.; Peter, M. E., Apoptosis signaling by death receptors. *Eur J Biochem* **1998**, *254* (3), 439–459.
27. Tait, S. W.; Green, D. R., Mitochondria and cell death: Outer membrane permeabilization and beyond. *Nat Rev Mol Cell Biol* **2010**, *11* (9), 621–632.
28. Green, D. R., The coming decade of cell death research: Five riddles. *Cell* **2019**, *177* (5), 1094–1107.
29. Edlich, F.; Banerjee, S.; Suzuki, M.; Cleland, M. M.; Arnoult, D.; Wang, C.; Neutzner, A.; Tjandra, N.; Youle, R. J., Bcl-x(L) retrotranslocates Bax from the mitochondria into the cytosol. *Cell* **2011**, *145* (1), 104–116.
30. Schellenberg, B.; Wang, P.; Keeble, J. A.; Rodriguez-Enriquez, R.; Walker, S.; Owens, T. W.; Foster, F.; Tanianis-Hughes, J.; Brennan, K.; Streuli, C. H.; Gilmore, A. P., Bax exists in a dynamic equilibrium between the cytosol and mitochondria to control apoptotic priming. *Mol Cell* **2013**, *49* (5), 959–971.
31. Todt, F.; Cakir, Z.; Reichenbach, F.; Emschermann, F.; Lauterwasser, J.; Kaiser, A.; Ichim, G.; Tait, S. W.; Frank, S.; Langer, H. F.; Edlich, F., Differential retrotranslocation of mitochondrial Bax and Bak. *Embo J* **2015**, *34* (1), 67–80.
32. Garner, T. P.; Reyna, D. E.; Priyadarshi, A.; Chen, H. C.; Li, S.; Wu, Y.; Ganesan, Y. T.; Malashkevich, V. N.; Cheng, E. H.; Gavathiotis, E., An autoinhibited dimeric form of BAX regulates the BAX activation pathway. *Mol Cell* **2016**, *63* (3), 485–497.
33. Galluzzi, L.; López-Soto, A.; Kumar, S.; Kroemer, G., Caspases connect cell-death signaling to organismal homeostasis. *Immunity* **2016**, *44* (2), 221–231.
34. Huber, K. L.; Serrano, B. P.; Hardy, J. A., Caspase-9 CARD: Core domain interactions require a properly formed active site. *Biochem J* **2018**, *475* (6), 1177–1196.
35. Riley, J. S.; Malik, A.; Holohan, C.; Longley, D. B., DED or alive: Assembly and regulation of the death effector domain complexes. *Cell Death Dis* **2015**, *6* (8), e1866.
36. Majno, G.; Joris, I., Apoptosis, oncosis, and necrosis. An overview of cell death. *Am J Pathol* **1995**, *146* (1), 3–15.
37. Wegner, K. W.; Saleh, D.; Degterev, A., Complex pathologic roles of RIPK1 and RIPK3: Moving beyond necroptosis. *Trends Pharmacol Sci* **2017**, *38* (3), 202–225.
38. Green, D. R., The coming decade of cell death research: Five riddles. *Cell* **2019**, *177* (5), 1094–1107.
39. Hitomi, J.; Christofferson, D. E.; Ng, A.; Yao, J.; Degterev, A.; Xavier, R. J.; Yuan, J., Identification of a molecular signaling network that regulates a cellular necrotic cell death pathway. *Cell* **2008**, *135* (7), 1311–1323.
40. Dondelinger, Y.; Declercq, W.; Montessuit, S.; Roelandt, R.; Goncalves, A.; Bruggeman, I.; Hulpiau, P.; Weber, K.; Sehon, C. A.; Marquis, R. W.; Bertin, J.; Gough, P. J.; Savvides, S.; Martinou, J. C.; Bertrand, M. J.; Vandenabeele, P., MLKL compromises plasma membrane integrity by binding to phosphatidylinositol phosphates. *Cell Rep* **2014**, *7* (4), 971–981.
41. Broz, P.; Pelegrín, P.; Shao, F., The gasdermins, a protein family executing cell death and inflammation. *Nat Rev Immunol* **2020**, *20* (3), 143–157.
42. Liu, X.; Xia, S.; Zhang, Z.; Wu, H.; Lieberman, J., Channelling inflammation: Gasdermins in physiology and disease. *Nat Rev Drug Discov* **2021**, *20* (5), 384–405.
43. Shi, J.; Gao, W.; Shao, F., Pyroptosis: Gasdermin-mediated programmed necrotic cell death. *Trends Biochem Sci* **2017**, *42* (4), 245–254.
44. Kayagaki, N.; Stowe, I. B.; Lee, B. L.; O'Rourke, K.; Anderson, K.; Warming, S.; Cuellar, T.; Haley, B.; Roose-Girma, M.; Phung, Q. T.; Liu, P. S.; Lill, J. R.; Li, H.; Wu, J.; Kummerfeld, S.; Zhang, J.; Lee, W. P.; Snipas, S. J.; Salvesen, G. S.; Morris, L. X.; Fitzgerald, L.; Zhang, Y.; Bertram, E. M.; Goodnow, C. C.; Dixit, V. M., Caspase-11 cleaves gasdermin D for non-canonical inflammasome signalling. *Nature* **2015**, *526* (7575), 666–671.

45. Shi, J.; Zhao, Y.; Wang, Y.; Gao, W.; Ding, J.; Li, P.; Hu, L.; Shao, F., Inflammatory caspases are innate immune receptors for intracellular LPS. *Nature* **2014**, *514* (7521), 187–192.
46. Tsuchiya, K.; Nakajima, S.; Hosojima, S.; Thi Nguyen, D.; Hattori, T.; Manh Le, T.; Hori, O.; Mahib, M. R.; Yamaguchi, Y.; Miura, M.; Kinoshita, T.; Kushiyama, H.; Sakurai, M.; Shiroishi, T.; Suda, T., Caspase-1 initiates apoptosis in the absence of gasdermin D. *Nat Commun* **2019**, *10* (1), 2091.
47. Wang, Y.; Gao, W.; Shi, X.; Ding, J.; Liu, W.; He, H.; Wang, K.; Shao, F., Chemotherapy drugs induce pyroptosis through caspase-3 cleavage of a gasdermin. *Nature* **2017**, *547* (7661), 99–103.
48. Santa Cruz Garcia, A. B.; Schnur, K. P.; Malik, A. B.; Mo, G. C. H., Gasdermin D pores are dynamically regulated by local phosphoinositide circuitry. *Nat Commun* **2022**, *13* (1), 52.
49. Wang, H.; Vilela, M.; Winkler, A.; Tarnawski, M.; Schlichting, I.; Yumerefendi, H.; Kuhlman, B.; Liu, R.; Danuser, G.; Hahn, K. M., LOVTRAP: An optogenetic system for photoinduced protein dissociation. *Nat Methods* **2016**, *13* (9), 755–758.
50. Tan, P.; He, L.; Han, G.; Zhou, Y., Optogenetic immunomodulation: Shedding light on antitumor immunity. *Trends Biotechnol* **2017**, *35* (3), 215–226.
51. England, C. G.; Ehlerding, E. B.; Cai, W., NanoLuc: A small luciferase is brightening up the field of bioluminescence. *Bioconjug Chem* **2016**, *27* (5), 1175–1187.
52. Wang, Q.; Wang, Y.; Ding, J.; Wang, C.; Zhou, X.; Gao, W.; Huang, H.; Shao, F.; Liu, Z., A bioorthogonal system reveals antitumour immune function of pyroptosis. *Nature* **2020**, *579* (7799), 421–426.
53. Chen, R.; Jing, J.; Siwko, S.; Huang, Y.; Zhou, Y., Intelligent cell-based therapies for cancer and autoimmune disorders. *Curr Opin Biotechnol* **2020**, *66*, 207–216.
54. Hong, M.; Clubb, J. D.; Chen, Y. Y., Engineering CAR-T cells for next-generation cancer therapy. *Cancer Cell* **2020**, *38* (4), 473–488.
55. Schubert, M. L.; Schmitt, M.; Wang, L.; Ramos, C. A.; Jordan, K.; Müller-Tidow, C.; Dreger, P., Side-effect management of chimeric antigen receptor (CAR) T-cell therapy. *Ann Oncol* **2021**, *32* (1), 34–48.

7 Control of Protein Levels in *Saccharomyces cerevisiae* by Optogenetic Modules that Act on Protein Synthesis and Stability

Sophia Hasenjäger, Jonathan Trauth, and Christof Taxis

CONTENTS

7.1 Introduction 99
7.2 Optogenetic Regulation of Protein Stability in *S. cerevisiae* 100
7.3 Light-Sensitive Protein Synthesis in Budding Yeast 101
7.4 Synergistic Optogenetic Multistep Control of Protein Levels 103
  7.4.1 A Simple Two-Step Protocol to Implement Synergistic Optogenetic Control of Target Protein Levels 104
  7.4.2 Optogenetic Control of Erg9 and Gdh1 105
7.5 Perspectives 106
Acknowledgments 107
References 107

7.1 INTRODUCTION

Optogenetic tools have revolutionized experimental approaches in neuroscience, but have applications in other biomedical research fields as well.[1,2] Many cellular processes are amenable for optogenetic regulation, namely protein synthesis, transport, stability, or enzymatic activity.[3] In general, optogenetic tools have a dual functionality: light perception and an effector domain that translates the light signal into a molecular response. Light sensing is achieved by diverse cofactors that are bound to the photoreceptor or the amino acid tryptophan that directly perceives ultra-violet B (UV-B) light. The activation of the light receptor leads to structural changes in the photoreceptor that are transmitted into a cellular signal by the effector domain.[4,5]

DOI: 10.1201/b22823-7

Microorganisms are easily accessible for optogenetic approaches due to the facile way light can be applied to them. Recently, budding yeast has been used intensively to generate optogenetic tools.[6,7] The main focus of yeast researchers lies in the development and usage of tools for biotechnology and cellular applications.[8] This chapter focuses on the implementation and usage of optogenetic tools in yeast to achieve efficient regulation of protein abundance by concomitant control of protein synthesis and protein stability in *Saccharomyces cerevisiae.*

7.2 OPTOGENETIC REGULATION OF PROTEIN STABILITY IN *S. CEREVISIAE*

Regulation of protein stability is one of the fundamental cellular principles to control protein activity, which is especially important for eukaryotic cells.[9] One of the main players in intracellular protein degradation is the ubiquitin-proteasome system that is involved in the degradation of the majority of cytosolic proteins.[10] Recognition of substrate proteins is provided by a degradation sequence (degron) that is recognized by the degradation machinery.[11,12] Several optogenetic tools have been developed to regulate protein stability via the ubiquitin-proteasome system. The method developed first relied on the presentation of a synthetic degradation sequence (degron) cODC1 (C-terminal sequence of murine ornithine decarboxylase) by the light-oxygen-voltage domain 2 (LOV2) of *Arabidopsis thaliana* through blue light-induced conformational changes within the domain.[13] The method was developed in *Saccharomyces cerevisiae*, and the same construct was used successfully in *Caenorhabditis elegans*. Similar constructs have been developed for use with mammalian cells and zebrafish.[14–16] Moreover, light-activated chemical inducers have been developed and tested in budding yeast, as well as in cell culture.[17,18]

The first implementation, also called photo-sensitive degron (psd), has been improved through several rounds of construct development, which resulted in a module with increased versatility and usability.[13,19–21] The half-lives of the different constructs fused to a stable red fluorescent protein have been measured between 3 and 20 min (Table 7.1). Yet it is not guaranteed that the same half-lives are measured for a specific yeast protein modified with a psd tag. An example of such behavior is the endoplasmic reticulum (ER) membrane protein Sec62 that is part of the ER protein import machinery.[22] Sec62-psd exhibited very fast degradation upon blue-light illumination and a half-time of 5 min, although the first generation psd module was used. Analysis of Sec62-psd degradation revealed that it was recognized by ER quality control mechanisms and degraded by an ER-associated protein degradation pathway, whereas soluble cytosolic proteins tagged with psd are degraded differently.[23]

Generally, we do not recommend including a red fluorescent tag in a construct due to incomplete degradation of fluorescent proteins through proteasomal degradation,[25] which could decrease the efficiency of target protein degradation. Usually, we use a tag that includes a triple-Myc (3myc) tag between the psd variant and the target protein for detection (Figure 7.1). The 3myc tag might also act as an inert linker sequence between the target protein and the psd module.

A detailed description of gene fusions with the psd module at the chromosomal level in *S. cerevisiae* has been published previously. The procedure can be performed

TABLE 7.1
Overview of the Different Photo-Sensitive Degron (psd) Modules

Construct	Half-life in RFP-psd fusion in blue light	Tagging plasmid with 3myc tag	Reference
psd	20	pDS96	[13]
psd$^{K92R\ E132A\ E155G}$	12	pDS170	[24]
psd3	5	pSH26	[20]
psd3^{V416L}	3	pDS325	[21]

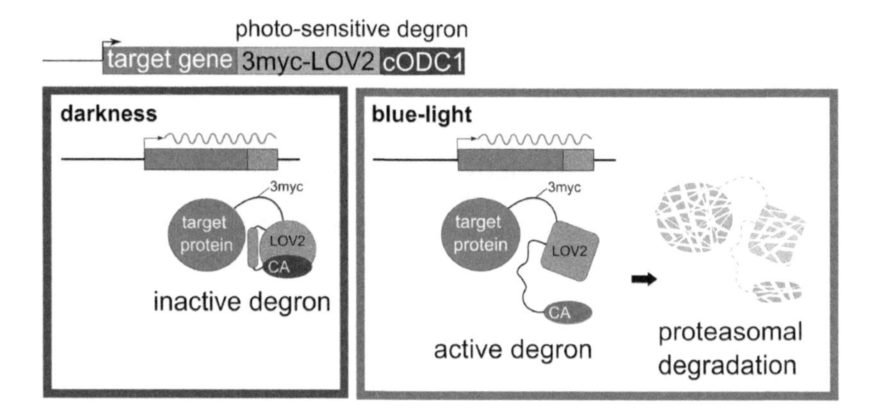

FIGURE 7.1 Scheme depicting the tagging procedure using a psd module to generate a light-sensitive mutant. The target gene is modified at the 3' end with the psd module to generate light sensitive target proteins.

similarly with newer psd modules not mentioned in the article.[24] We generated diverse light-sensitive mutants of essential genes using this polymerase chain reaction (PCR)-based tagging procedure for the characterization of unknown essential proteins and to characterize a protein degradation pathway.[13,19–21,23,26,27]

7.3 LIGHT-SENSITIVE PROTEIN SYNTHESIS IN BUDDING YEAST

Many tools exist that focus on light-activated transcription in budding yeast or other organisms,[7,28,29] whereas the opposite, photo-inhibited transcription, has been somewhat neglected. A few examples exist that show light-induced reduction of gene expression in mammalian cell culture.[30,31] Additionally, the OptoINVRT constructs can be used to repress genes in response to light; they have been developed for *S. cerevisiae* based on the endogenous galactose-regulon for biotechnology applications.[32] The photo-sensitive transcription factor (psTF), a tool to induce target gene transcription in darkness and shut-down this activity upon blue-light illumination, is based solely on non-yeast-sequences to reduce the influence on the endogenous metabolism (Figure 7.2).[20]

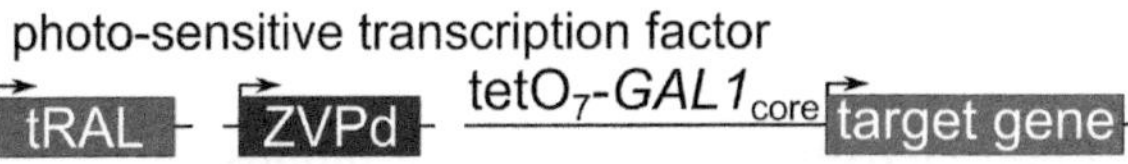

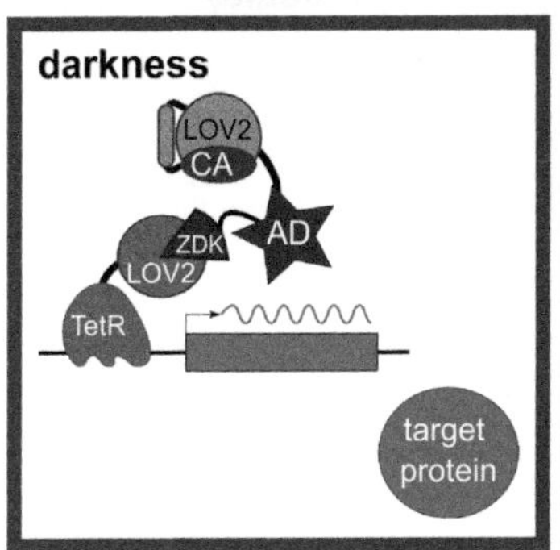

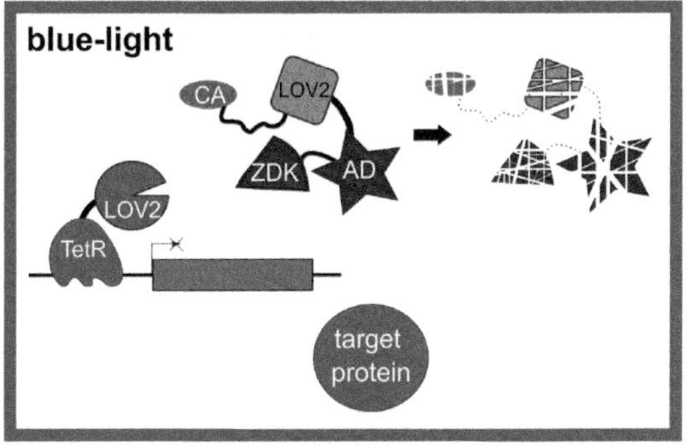

FIGURE 7.2 Scheme showing the architecture of the photosensitive transcription factor (psTF). Abbreviations: tRAL: tetR-*As*LOV2; ZVPd: Zdk1-VP16–3myc-psd3; tetR: tetracycline resistance Repressor; $tetO_7$: seven tetR operators; *GAL1*: galactose metabolism gene 1; LOV2: Avena sativa light-oxygen-voltage domain 2; ZDK: Zdark 1; AD: activation domain; CA: cODC1 degron.

TABLE 7.2
Overview of the Different psTF Variants (Shown in Decreasing Light-Sensitivity)

As LOV2 variant	Dark-state reversion time [sec][33]	Plasmid name[21]
*As*LOV2^{V416L}	496 ± 38	pJG22
*As*LOV2^{V416I}	239 ± 5	pJG21
*As*LOV2	18.5 ± 3.7	pDS242
*As*LOV2^{V416T}	5.0 ± 2.0	pJG23
*As*LOV2^{I427T}	1.7 ± 0.6	pJG24

The psTF is based on the photosensitive interaction between the *Avena sativa* LOV2 (*As*LOV2) domain and the synthetic Zdark1 peptide.[33] The so-called LOV2 trap and release of protein (LOVTRAP) system was developed as a dark-induced protein-protein interaction tool by the Hahn group. The psTF was generated by fusing the LOV2 domain with the DNA-binding domain tetR and fusing the peptide Zdark1 with VP16.[20] Variation of the transcription factor characteristics and the light response has been achieved with mutational influence on the dark-reversion kinetics of the *As*LOV2 domain using the well-documented mutations V416I, V416L, V416T, and I427T.[21] The mutations V416L and V416I lead to the higher light sensitivity of the psTF, whereas the mutations V416T and I427T reduce the light sensitivity (Table 7.2).

The differences in the psTF variants are the result of changes in the dark-state reversion of the mutated *As*LOV2 photoreceptors. The variants with higher light sensitivity have a prolonged time of dark-state reversion, whereas the less sensitive ones have a shortened reversion compared to the *As*LOV2 domain.[33] Transcriptional control by the psTF has so far only been used in synthetic circuits, but the achievable light-regulation combined with the mutational variation in light-response promises broad applicability in yeast or other eukaryotic microorganisms for both biotechnology applications and basic research (see section 7.4).

7.4 SYNERGISTIC OPTOGENETIC MULTISTEP CONTROL OF PROTEIN LEVELS

Synergistic regulation of protein levels by a regulatory network enhances the switchability of protein abundance by using a light signal that acts concomitantly on more than one effector. So far, the implemented networks control transcription and protein stability simultaneously. Three different networks have been generated for mammalian cell culture and *S. cerevisiae*.[21,34,35] In budding yeast, the psd module has been combined with the psTF to achieve very high light/dark switching ratios (Figure 7.3).[20,21] Upon light-illumination, the off-rate kinetics is controlled by target protein degradation evoked by the psd module, whereas the psTF ensures the termination of target protein synthesis.[21]

Overall, dual regulation of protein synthesis and target protein stability ensures rapid and efficient inactivation of the target.

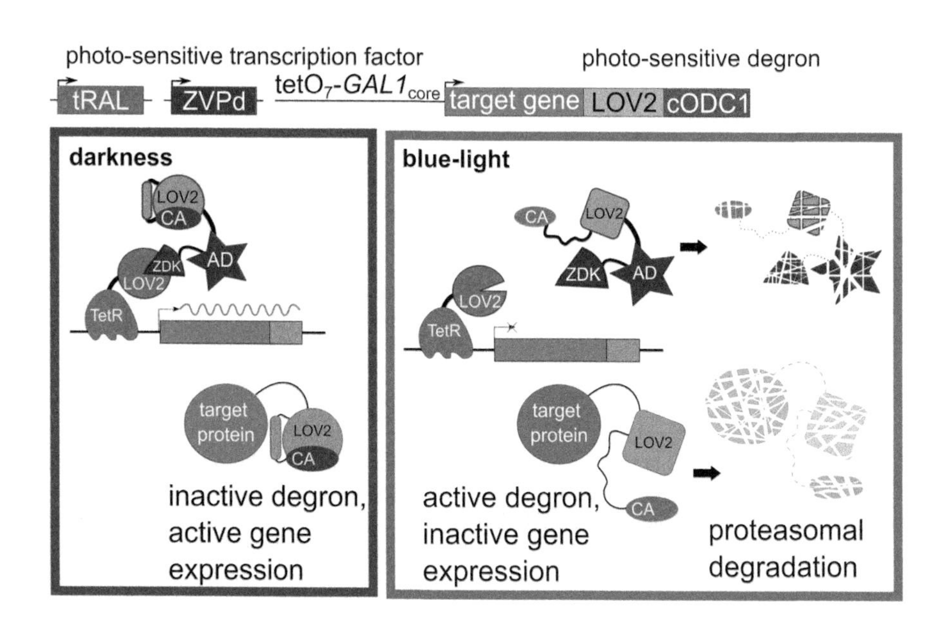

FIGURE 7.3 Synergistic regulation of protein levels comprises gene expression with the photosensitive transcription factor and control of protein stability with a psd module.

7.4.1 A Simple Two-Step Protocol to Implement Synergistic Optogenetic Control of Target Protein Levels

Generation of a construct for synergistic optogenetic multistep control (SOMCo) of an essential protein in yeast is easy and can be achieved following a simple two-step protocol. In the first step, the construct containing the target gene is assembled by homologous recombination in yeast using DNA fragments generated by PCR. In this way, all the necessary fragments are combined in one vector and different variants of psTF or psd modules can be exchanged to generate constructs with slightly different behaviors or using different tags for protein recognition (Figure 7.4).

Thus, the first step comprises the assembly and quality control of the vector with the regulatory sequences and the target gene. We used this approach for the assembly of several vectors. The example of oligonucleotide sequences for the assembly of a vector with *CDC20* as the target gene is given in Table 7.3. The fluorescent protein mCherry was included here to facilitate the recognition of correct clones by a flow cytometer; an alternative is the usage of one of the 3myc-psd variants shown in Table 7.1. Alternatives for the tetR-*As*LOV2 construct are given in Table 7.2. The fastest degradation of the target and highest light sensitivity of the transcription factor is expected with the usage of 3myc-psd3^{V416L} and tetR-*As*LOV2^{V416L}. The names of the oligos correspond to the labeling of DNA fragments in Figure 7.4.

Due to the orthogonal nature of psd module and psTF, the different variants that have been described in Sections 7.2 and 7.3 can be mixed freely. The modular approach to generating the complete SOMCo module in one plasmid enhances the

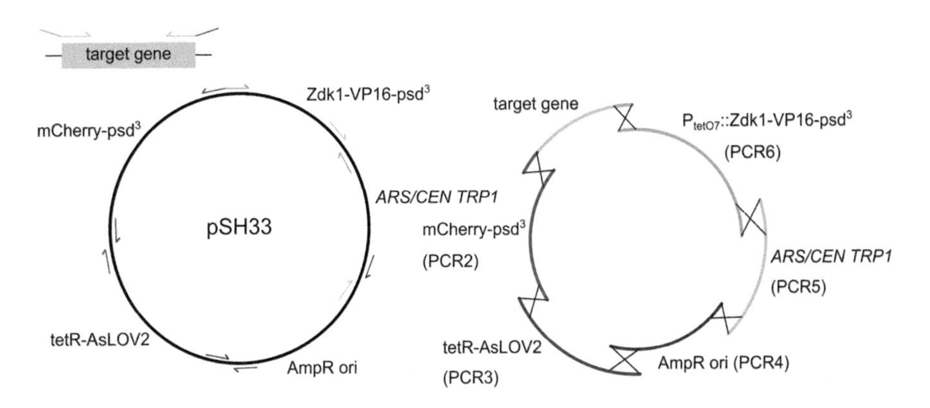

FIGURE 7.4 Scheme showing the general combination of psTF and psd modules with the target gene to generate a construct for synergistic optogenetic multistep control (SOMCo) of target protein abundance in yeast. Using cassettes for the auxotrophic marker of the vector and the replication type (*ARS/CEN* or 2µ) facilitates usage by variation of the selection marker and the plasmid copy number. Usage of 3'-extensions of the primer sequences allows customized composition of the final plasmid. Alternative constructs with different behavior require adaptation of primer sequences and additional templates: mCherry-psd3V416L: pDS309; usage of 3myc-psd variants: see Table 7.1; usage of tetR-AsLOV2 variants: see Table 7.2

TABLE 7.3
Oligonucleotide Sequences to Generate a Plasmid with *CDC20* as Target Gene

cdc20-down	GATGGCCATGTTATCCTCCTCGCCCTTGCTCACCATCCTGATCAAATATT GGCTGG
cdc20-uprev	AACGTCAAGGAGAAAAAACTATAACTAGTGGCCTATATGCCAGAAAGC TCTAGAGATAAG
PCR2for	TGGGCTTGATCCACCAACC
PCR2rev	ATGGTGAGCAAGGGCGAGG
PCR3for	TCTCCCCGCGCGTTGGC
PCR3rev	CCTATCCCACTAAAGGGAAC
PCR4for	ACCTGAGAGCAGGAAGAGC
PCR4rev	GCTCACTGCCCGCTTTCC
PCR5for	GTCTATTCTTTTGATTTATAAGGG
PCR5rev	CTCTAGGGGGATCGCCAAC
PCR6for	ATAGGCCACTAGTTATAGTTTTTTC
PCR6rev	CGTTAAATTTTTGTTAAATCAGCTC

flexibility of the whole approach (Figure 7.4). The result is tailored protein abundance adapted to the necessities of the application.

In the second step, the SOMCo vector is transformed into wild-type yeast cells. The chromosomal copy of the target gene is then deleted by a conventional knock-out cassette, e.g., generated by PCR as described previously.[36,37] An alternative is to do a CRISPR/Cas-based knock-out.[38] In the latter case, care has to be taken that the small guide RNA is not targeting a sequence that is present in the SOMCo vector with the target gene as well.

7.4.2 Optogenetic Control of Erg9 and Gdh1

We generated two strains with SOMCo modules to illustrate the feasibility of the approach towards cellular targets. We choose the non-essential gene *GDH1* and the essential gene *ERG9* as targets. We compared the regulation efficiency of the SOMCo approach with variants tagged with a 3myc or a psd3 tag (Figure 7.5). The non-modified variants of tetR-*As*LOV2 and psd3 were used for both constructs.

The Gdh1 levels were almost perfectly regulated by the SOMCo approach (Figure 7.5A). The Gdh1 abundance was nearly identical in the P_{tetO7}-*GDH1–3myc-psd3* construct in darkness compared to the *GDH1–3myc* strain. Blue-light illumination of the cells resulted in an almost complete disappearance of Gdh1. Thus, SOMCo regulation of *GDH1* results in a conditional knock-out.

The regulation or *ERG9* by a SOMCo construct was similarly successful, although the Erg9 levels in darkness are higher in the SOMCo construct (P_{tetO7}-*ERG9–3myc-psd3*) than in the *ERG9–3myc* strain. Blue-light illumination results in a profound reduction of Erg9 levels to roughly 30% of Erg9–3myc levels (Figure 7.5B). Overall, SOMCo regulation allows precise light control of cellular targets with a high switching ratio.

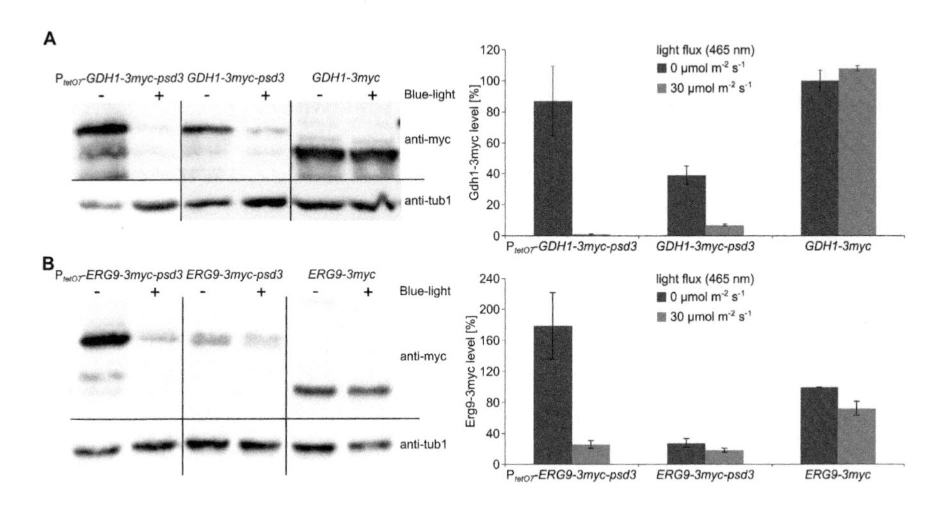

FIGURE 7.5 Synergistic optogenetic multistep control of Gdh1 and Erg9 compared with psd-dependent regulation and wild type protein levels. (A) Yeast strains with the SOMCo construct (P_{tetO7}-*GDH1–3myc-psd3*), Gdh1–3myc-psd3, and Gdh1–3myc were analyzed by immunoblotting, Cells were grown in darkness or in the presence of blue light (465 nm, 30 µmol m^{-2} s^{-1}). The *GDH1* SOMCo construct was used in a *GDH1* knockout strain. Left site: immunoblot analysis; right side: quantification of immunoblotting results (n=3; error bars: standard error of the mean). Please note that for quantification of signal intensities the samples were loaded on one gel to compare the abundance directly next to each other. One *GDH1*–3myc sample served as reference (100%) for the other samples. (B) Same as in **A** using *ERG9* as target gene.

7.5 PERSPECTIVES

The generation of SOMCo constructs is relatively easy and straightforward. Following a simple two-step protocol, as detailed previously, is sufficient to generate the necessary construct. One can expect a high switching ratio between dark and light states already with the wild-type modules for tetR-*As*LOV2 and psd3. In case the depletion rate is not sufficient, one can replace the modules with the V416L variants.

Yet, the *ERG9* target demonstrates an issue that might affect the application: increased or decreased expression of the target in the dark state. A possible solution for such issues is to change the core promoter of the $tetO_7$ promoter. The described construct uses the *GAL1* core promoter. It has been shown recently that variations in the core promoter sequence or the number of transcription factor binding sites allows modulation of light-regulated transcription.[39] In addition, a decrease in target gene expression could also be achieved by using very low amounts of light instead of keeping the cells in darkness.[13,20,21] Such adaptations will result in tailored constructs and experimental conditions for optimized control of cell behavior and target protein abundance.

In summary, the SOMCo modules provide a generalized approach for rapid and robust switching of protein activity from an active state to an inactive one. Modules with different behaviors are available for target construct optimization. These versatile molecular tools can be tailored for biomedical research and biotechnology applications by taking advantage of the remarkable features of light as the stimulus.

ACKNOWLEDGMENTS

The authors thank Daniela Störmer for her excellent technical assistance. This work was supported by DFG grant TA320/7–1 and the BMBF grant 031B0358A. SH acknowledges a MARA-stipend from the Philipps-University of Marburg.

REFERENCES

1. Manoilov, K. Y., Verkhusha, V. V., & Shcherbakova, D. M. (2021). A guide to the optogenetic regulation of endogenous molecules. *Nature Methods*, *18*(9), 1027–1037. https://doi.org/10.1038/s41592-021-01240-1
2. Gautier, A., Gauron, C., Volovitch, M., Bensimon, D., Jullien, L., & Vriz, S. (2014). How to control proteins with light in living systems. *Nature Chemical Biology*, *10*(7), 533–541. https://doi.org/10.1038/nchembio.1534
3. Rost, B. R., Schneider-Warme, F., Schmitz, D., & Hegemann, P. (2017). Optogenetic tools for subcellular applications in neuroscience. *Neuron*, *96*(3), 572–603. https://doi.org/10.1016/j.neuron.2017.09.047
4. Mathes, T. (2016). Natural resources for optogenetic tools. *Methods in Molecular Biology (Clifton, N.J.)*, *1408*, 19–36. https://doi.org/10.1007/978-1-4939-3512-3_2
5. Yin, R., & Ulm, R. (2017). How plants cope with UV-B: From perception to response. *Current Opinion in Plant Biology*, *37*, 42–48. https://doi.org/10.1016/J.PBI.2017.03.013
6. Salinas, F., Rojas, V., Delgado, V., Agosin, E., & Larrondo, L. F. (2017). Optogenetic switches for light-controlled gene expression in yeast. *Applied Microbiology and Biotechnology*, *101*, 2629–2640. https://doi.org/10.1007/s00253-017-8178-8
7. Pérez, A. L. A., Piva, L. C., Fulber, J. P. C., de Moraes, L. M. P., De Marco, J. L., Vieira, H. L. A., Coelho, C. M., Reis, V. C. B., & Torres, F. A. G. (2021). Optogenetic strategies for the control of gene expression in yeasts. *Biotechnology Advances*, *54*, 107839. https://doi.org/10.1016/J.BIOTECHADV.2021.107839
8. Figueroa, D., Rojas, V., Romero, A., Larrondo, L. F., & Salinas, F. (2021). The rise and shine of yeast optogenetics. *Yeast*, *38*(2), 131–146. https://doi.org/10.1002/YEA.3529
9. Hershko, A., & Ciechanover, A. (1998). The ubiquitin system. *Annual Review of Biochemistry*, *67*, 425–479. https://doi.org/10.1007/978-1-4899-1922-9_1
10. Kwon, Y. T., & Ciechanover, A. (2017). The ubiquitin code in the ubiquitin-proteasome system and autophagy. *Trends in Biochemical Sciences*, *42*, 873–886. https://doi.org/10.1016/j.tibs.2017.09.002
11. Trauth, J., Scheffer, J., Hasenjäger, S., & Taxis, C. (2019). Synthetic control of protein degradation during cell proliferation and developmental processes. *ACS Omega*, *4*(2). https://doi.org/10.1021/acsomega.8b03011
12. Trauth, J., Scheffer, J., Hasenjäger, S., & Taxis, C. (2020). Strategies to investigate protein turnover with fluorescent protein reporters in eukaryotic organisms. *AIMS Biophysics*, *7*(2), 90–118. https://doi.org/10.3934/biophy.2020008
13. Renicke, C., Schuster, D., Usherenko, S., Essen, L. O., & Taxis, C. (2013). A LOV2 domain-based optogenetic tool to control protein degradation and cellular function. *Chemistry and Biology*, *20*(4), 619–626. https://doi.org/10.1016/j.chembiol.2013.03.005
14. Bonger, K. M., Rakhit, R., Payumo, A. Y., Chen, J. K., & Wandless, T. J. (2014). General method for regulating protein stability with light. *ACS Chemical Biology*, *9*(1), 111–115. https://doi.org/10.1021/cb400755b

15. Hermann, A., Liewald, J. F., & Gottschalk, A. (2015). A photosensitive degron enables acute light-induced protein degradation in the nervous system. *Current Biology*, *25*(17), R749–R750. https://doi.org/10.1016/j.cub.2015.07.040
16. Sun, W., Zhang, W., Zhang, C., Mao, M., Zhao, Y., Chen, X., & Yang, Y. (2017). Light-induced protein degradation in human-derived cells. *Biochemical and Biophysical Research Communications*, *487*(2), 241–246. https://doi.org/10.1016/j.bbrc.2017.04.041
17. Delacour, Q., Li, C., Plamont, M. A., Billon-Denis, E., Aujard, I., Le Saux, T., Jullien, L., & Gautier, A. (2015). Light-activated proteolysis for the spatiotemporal control of proteins. *ACS Chemical Biology*, *10*(7), 1643–1647. https://doi.org/10.1021/acschembio.5b00069
18. Reynders, M., Matsuura, B. S., Bérouti, M., Simoneschi, D., Marzio, A., Pagano, M., & Trauner, D. (2020). PHOTACs enable optical control of protein degradation. *Science Advances*, *6*(8), eaay5064. https://doi.org/10.1126/sciadv.aay5064
19. Usherenko, S., Stibbe, H., Muscò, M., Essen, L. O., Kostina, E. A., & Taxis, C. (2014). Photo-sensitive degron variants for tuning protein stability by light. *BMC Systems Biology*, *8*, 128. https://doi.org/10.1186/s12918-014-0128-9
20. Hasenjäger, S., Trauth, J., Hepp, S., Goenrich, J., Essen, L.-O., & Taxis, C. (2019). Optogenetic downregulation of protein levels with an ultrasensitive switch. *ACS Synthetic Biology*, *8*(5), 1026–1036. https://doi.org/10.1021/acssynbio.8b00471
21. Pook, B., Goenrich, J., Hasenjäger, S., Essen, L.-O., Spadaccini, R., & Taxis, C. (2021). An optogenetic toolbox for synergistic regulation of protein abundance. *ACS Synthetic Biology*, acssynbio.1c00350. https://doi.org/10.1021/ACSSYNBIO.1C00350
22. Lyman, S. K., & Schekman, R. (1997). Binding of secretory precursor polypeptides to a translocon subcomplex is regulated by BiP. *Cell*, *88*(1), 85–96. https://doi.org/10.1016/S0092-8674(00)81861-9
23. Scheffer, J., Hasenjäger, S., & Taxis, C. (2019). Degradation of integral membrane proteins modified with the photosensitive degron module requires the cytosolic endoplasmic reticulum—associated degradation pathway. *Molecular Biology of the Cell*, *30*(20), 2558–2570. https://doi.org/10.1091/mbc.E18-12-0754
24. Lutz, A. P., Renicke, C., & Taxis, C. (2016). Controlling protein activity and degradation using blue light. In Kianianmomeni A. (Ed.), *Methods in Molecular Biology* (Vol. 1408, pp. 67–78). Humana Press. https://doi.org/10.1007/978-1-4939-3512-3_5
25. Khmelinskii, A., Meurer, M., Ho, C.-T. C.-T., Besenbeck, B., Fuller, J., Lemberg, M. K., Bukau, B., Mogk, A., & Knop, M. (2016). Incomplete proteasomal degradation of green fluorescent proteins in the context of tandem fluorescent protein timers. *Molecular Biology of the Cell*, *27*(2), 360–370. https://doi.org/\url{10.1091/mbc.E15-07-0525}
26. Paul, V. D., Mühlenhoff, U., Stümpfig, M., Seebacher, J., Kugler, K. G., Renicke, C., Taxis, C., Gavin, A.-C., Pierik, A. J., & Lill, R. (2015). The deca-GX<inf>3</inf> proteins Yae1-Ltol function as adaptors recruiting the ABC protein Rli1 for iron-sulfur cluster insertion. *ELife*, *4*(JULY2015). https://doi.org/10.7554/eLife.08231
27. Lutz, A. P., Schladebeck, S., Renicke, C., Spadaccini, R., Mösch, H. U., & Taxis, C. (2018). Proteasome activity is influenced by the HECT_2 protein Ipa1 in budding yeast. *Genetics*, *209*(1), 157–171. https://doi.org/10.1534/genetics.118.300744
28. Baumschlager, A., & Khammash, M. (2021). Synthetic biological approaches for optogenetics and tools for transcriptional light-control in bacteria. *Advanced Biology*, *5*(5). https://doi.org/10.1002/ADBI.202000256
29. Yamada, M., Nagasaki, S. C., Ozawa, T., & Imayoshi, I. (2020). Light-mediated control of Gene expression in mammalian cells. *Neuroscience Research*, *152*, 66–77. https://doi.org/10.1016/J.NEURES.2019.12.018
30. Pathak, G. P., Spiltoir, J. I., Höglund, C., Polstein, L. R., Heine-Koskinen, S., Gersbach, C. A., Rossi, J., & Tucker, C. L. (2017). Bidirectional approaches for optogenetic

regulation of gene expression in mammalian cells using Arabidopsis cryptochrome 2. *Nucleic Acids Research*, *45*(20), e167. https://doi.org/10.1093/nar/gkx260

31. Konermann, S., Brigham, M. D., Trevino, A. E., Hsu, P. D., Heidenreich, M., Cong, L., Platt, R. J., Scott, D. A., Church, G. M., & Zhang, F. (2013). Optical control of mammalian endogenous transcription and epigenetic states. *Nature*, *500*(7463), 472–476. https://doi.org/10.1038/NATURE12466
32. Zhao, E. M., Zhang, Y., Mehl, J., Park, H., Lalwani, M. A., Toettcher, J. E., & Avalos, J. L. (2018). Optogenetic regulation of engineered cellular metabolism for microbial chemical production. *Nature*, *555*(7698), 683–687. https://doi.org/10.1038/nature26141
33. Wang, H., Vilela, M., Winkler, A., Tarnawski, M., Schlichting, I., Yumerefendi, H., Kuhlman, B., Liu, R., Danuser, G., & Hahn, K. M. (2016). LOVTRAP: An optogenetic system for photoinduced protein dissociation. *Nature Methods*, *13*(9), 755–758. https://doi.org/10.1038/nmeth.3926
34. Pathak, G. P., Spiltoir, J. I., Höglund, C., Polstein, L. R., Heine-Koskinen, S., Gersbach, C. A., Rossi, J., & Tucker, C. L. (2017). Bidirectional approaches for optogenetic regulation of gene expression in mammalian cells using Arabidopsis cryptochrome 2. *Nucleic Acids Research*, *45*(20), e167–e167. https://doi.org/10.1093/NAR/GKX260
35. Baaske, J., Gonschorek, P., Engesser, R., Dominguez-Monedero, A., Raute, K., Fischbach, P., Müller, K., Cachat, E., Schamel, W. W. A., Minguet, S., Davies, J. A., Timmer, J., Weber, W., & Zurbriggen, M. D. (2018). Dual-controlled optogenetic system for the rapid down-regulation of protein levels in mammalian cells. *Scientific Reports*, *8*(1), 15024. https://doi.org/10.1038/s41598-018-32929-7
36. Wach, A., Brachat, A., Pöhlmann, R., & Philippsen, P. (1994). New heterologous modules for classical or PCR-based gene disruptions in Saccharomyces cerevisiae. *Undefined*, *10*(13), 1793–1808. https://doi.org/10.1002/YEA.320101310
37. Janke, C., Magiera, M. M., Rathfelder, N., Taxis, C., Reber, Sa., Maekawa, H., Moreno-Borchart, A., Doenges, G., Schwob, E., Schiebel, E., & Knop, M. (2004). A versatile toolbox for PCR-based tagging of yeast genes: New fluorescent proteins, more markers and promoter substitution cassettes. *Yeast*, *21*(11), 947–962. https://doi.org/10.1002/yea.1142
38. Mans, R., van Rossum, H. M., Wijsman, M., Backx, A., Kuijpers, N. G. A., van den Broek, M., Daran-Lapujade, P., Pronk, J. T., van Maris, A. J. A., & Daran, J. M. G. (2015). CRISPR/Cas9: A molecular Swiss army knife for simultaneous introduction of multiple genetic modifications in Saccharomyces cerevisiae. *FEMS Yeast Research*, *15*(2), 1–15. https://doi.org/10.1093/femsyr/fov004
39. Benzinger, D., & Khammash, M. (2018). Pulsatile inputs achieve tunable attenuation of gene expression variability and graded multi-gene regulation. *Nature Communications*, *9*(1). https://doi.org/10.1038/s41467-018-05882-2

8 Optogenetics as a Tool to Study Neurodegeneration and Signal Transduction

Prabhat Tiwari and Nicholas S. Tolwinski

CONTENTS

8.1 Introduction 111
8.2 Commonly Used Non-Opsin Photosensitive Proteins 112
8.3 Using Optogenetics to Study Protein Aggregation-Related Diseases 112
8.4 Using Optogenetics to Study Signalling Pathways 114
8.5 CRY2 Dependent Optogenetic System 114
 8.5.1 Activators 114
 8.5.2 Inhibitors 114
8.6 LOV-Based Optogenetic System 116
 8.6.1 Opto-SOS 116
8.7 Optogentic Systems Based on Caging 117
 8.7.1 Opto-YAP 117
 8.7.2 DRONPA-Derived Photoswitchable Kinases 117
8.8 Optogenetics and Microscopy 117
8.9 Discussion 119
Acknowledgments 119
Reference 119

8.1 INTRODUCTION

Studying cell behaviour within tissues during development or adulthood requires techniques for genetic perturbation. This is often straightforward during early stages of development, but requires more complex approaches as organisms mature. Optogenetics is a technique that uses genetically engineered photosensitive molecules that can be expressed and activated in a spatiotemporal manner by the application of light. It can be applied to allow high precision modulation of cellular activity with subcellular resolution. The very first studies on optogenetics used a light-sensitive cation channel (channelrhodopsin) to modulate neural activity [1,2]. These tools were expanded with other opsin-based systems like halorhodopsin to study neuronal activity. The next step used various plant and bacterial photoreceptors to generate

DOI: 10.1201/b22823-8

non-opsin, light controllable protein systems leading to tremendous advances in photosensitive systems. These have been used to study cell division, cell migration, RNA binding, signal transduction and neurodegeneration, among many others[3–8]. In this chapter, we focus on the most common non-opsin photosensitive systems: Cryptochrome 2 (CRY2), Light oxygen voltage (LOV) and DRONPA. These can be applied in different ways based on their design and placement within pathways.

8.2 COMMONLY USED NON-OPSIN PHOTOSENSITIVE PROTEINS

CRY2 can be used in a variety of ways. For example, it can be induced by light to oligomerize, leading to clustering of the moiety associated with it, such as human amyloid-b tagged with CRY2[9]. CRY2 can also be used for directed protein-protein interaction and localization to specific subcellular compartments via its ability to heterodimerize with CIB[10].

LOV or Light-oxygen-voltage domain-based system can work through conformational change (AsLOV[11]), homodimerization (VfAU1-LOV[12]), heterodimerization (iLID[13]), and dissociation (LOVTRAP[14]). This variety allows for the design of many different engineered protein activators and inhibitors.

DRONPA is a photoswitchable fluorescent protein which can be switched off with 500 nm light and switched on with 400 nm light. While on, it can be used to cage a protein of interest thus inhibiting the protein's function. After exposure to 500 nm light, the protein is released to perform its function and can be inactivated upon exposure to 400 nm light as shown for the mitogen-activated protein kinase (MAPK) pathway through the caging of MEK[15].

8.3 USING OPTOGENETICS TO STUDY PROTEIN AGGREGATION-RELATED DISEASES

Neurodegeneration is a progressive loss of neurons and their function as exemplified by Alzheimer's disease (AD). AD is the most common cause of dementia accounting for about 60% of cases, resulting in progressive loss of memory and thinking ability, which eventually interferes with simple daily tasks. The brains of AD patients show extracellular accumulation of amyloid-β (Aβ) plaques[16]. Extracellular accumulation of amyloid-β plaques was thought to be the main driver in AD progression, but several studies suggest otherwise. Brains of non-symptomatic individuals show Aβ accumulation[17], and some cases of AD manifest the pathological features without extracellular Aβ accumulation[18–20]. There have been several studies reporting intracellular Aβ accumulation, which may have pathological effect irrespective of Aβ plaque formation[21,22]. Generally, human studies on intracellular Aβ are limited to correlative data, giving animal models a significant edge. The addition of optogenetics to these animal models adds an *in vivo* aspect to the study of Aβ in neurodegeneration. Models in *C. elegans* (nematode worm), *Drosophila melanogaster* (fruit fly) and *Danio rerio* (zebrafish) have been made to study the effect of Aβ accumulation and aggregation/oligomerization[9,23]. These animal models were made by using the modified version of an optogenetic protein from *Arabidopsis thaliana*, cryptochrome2 (CRY2, Figure 8.1). In presence of blue light (488nm) it oligomerizes quickly and this oligomerization can be reversed rapidly by turning off blue light. The photolyase homology region of the CRY2 protein was fused to mCherry fluorescent protein and used to follow the localization of the protein of interest. In the presence of blue light, Aβ-CRY2-mCherry showed aggregation, resulting in metabolic defects, loss of mitochondria, and inflammation in the worm and fruit fly, and leading to neuronal degeneration in the fruit fly and zebrafish models. Lifespan and behavioral defects were also observed in the worm and fruit fly models. Although CRY2-mCherry alone showed reversal of oligomerization[24], in the absence of blue light the aggregates of Aβ-CRY2 mCherry did not show reversal after oligomerization, likely due to Aβ aggregation. Taken together these optogenetic models provide an insight into the biology of Aβ by separating expression and aggregation of the protein[9]. A detailed protocol is available[23].

In order to ameliorate the effects imposed by amyloid, a common treatment (lithium) was used to inhibit the GSK3 kinase that activates Wnt signaling. The Wnt signaling pathway has been shown to play a role in the AD pathology and progression[25–27]. Addition of Li^+ rescued the life span reduction phenotype in fruit fly expressing Aβ-CRY2 mCherry[9]. This finding was further extended using Li^+ to

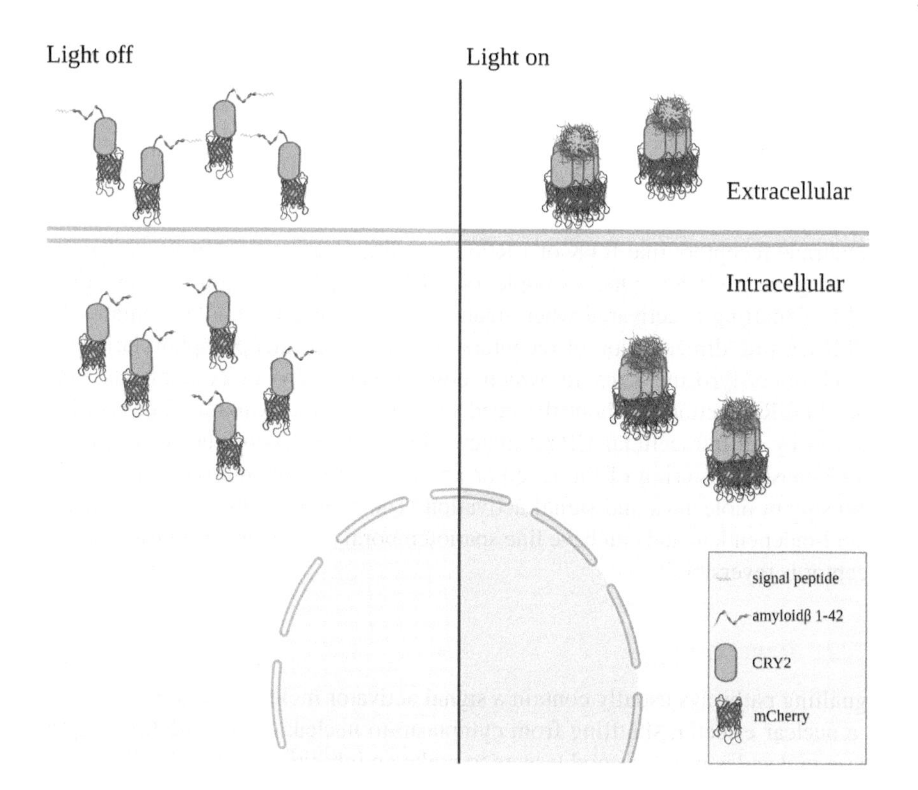

FIGURE 8.1 Opto-Aβ. The Cry2-Aβ-mCherry construct is expressed in the cytoplasm and does not aggregate in the absence of light. In the presence of blue light (488nm), CRY2 mediates clustering of Aβ causing oligomer, aggregate or plaque formation. Addition of a signal peptide to Cry2-Aβ-mCherry would lead to secretion and induction of clustering and plaque formation in the extracellular space upon blue light activation.

Wnt signaling directly by genetic means as this drug affects a variety of targets. These experiments were conducted in *Drosophila* adult gut stem cells to model the effect of Wnt on Aβ accumulation and aggregation[28]. Wnt expression rescued lifespan and tissue degeneration induced by Aβ-CRY2 mCherry[28], suggesting that the optogenetic model could be used to study treatment of Aβ in neurodegeneration.

8.4 USING OPTOGENETICS TO STUDY SIGNALLING PATHWAYS

Signal transduction is a tightly controlled phenomenon regulating the development and growth of an animal[29]. The spatiotemporal pattern of pathway activation is important for a variety of different cellular processes like translocation of transcription factors and dynamics of membrane and cytoskeleton, among many others[30]. As cells within tissues communicate through signals, optogenetics allows for spatiotemporal activation of signaling pathways. In this section, we will review a few recent examples of optogenetic regulation of signalling pathways based on LOV, CRY2, and DRONPA proteins.

8.5 CRY2 DEPENDENT OPTOGENETIC SYSTEM

The CRY2 system has been used to induce oligomerization of the tagged protein. This system thus can work as an activator or inhibitor of a signalling pathway depending on the protein it is fused to.

8.5.1 Activators

Membrane receptors like RTK or LRP6 need dimerization to activate downstream signaling events[24,31,32]. One example of CRY2-based activation is Opto-EGFR. EGFR signalling is activated when ligand binds to the extracellular domain of the EGFR causing dimerization of receptors and inducing trans-phosphorylation making phosphor-tyrosine sites to which downstream pathway components bind[33]. Opto-EGFR functions without the need for extracellular ligand as dimerization is induced by an intracellular CRY2 moiety (Figure 8.2). Upon blue light induction, CRY2 causes clustering of the receptor and thus activation and recruitment of the downstream molecules and signal activation. The benefit of this system is that it is ligand-independent and can have fine spatiotemporal control, as the clustering of the receptor is reversible[34].

8.5.2 Inhibitors

Signalling pathways usually contain a signal activator molecule which can function as a nuclear effector, shuttling from cytoplasm to nucleus to activate transcription. These molecules provide good targets to make an inhibition switch. In RTK signalling, the kinase ERK (MAPK) is phosphorylated and translocates to the nucleus where it phosphorylates transcription factors. An ERK molecule tagged with CRY2 can be aggregated in presence of blue light and is unable to translocate to nucleus thus acting as an inhibition to RTK signalling pathway (Figure 8.3)[34]. This principle

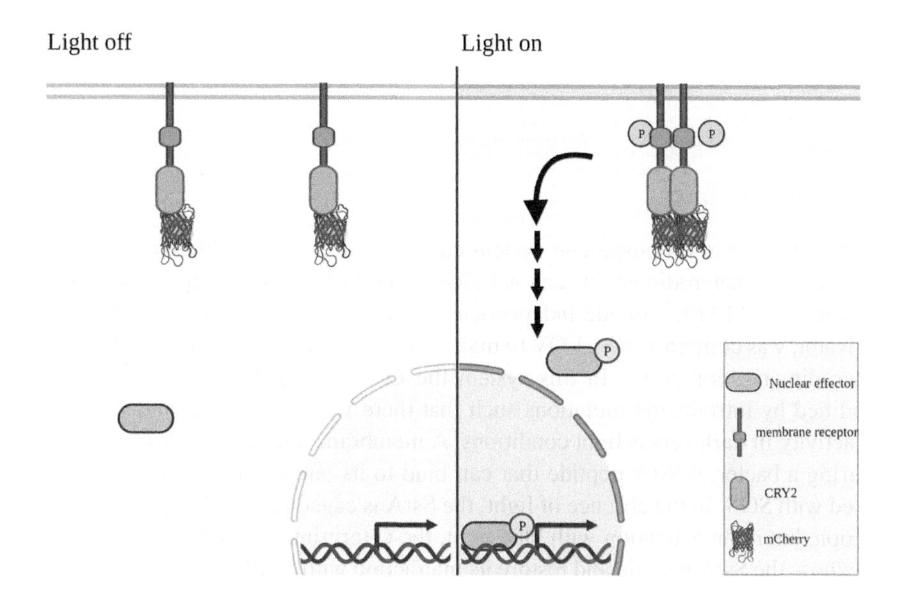

FIGURE 8.2 CRY-based signal activation. Membrane receptors activated via dimerization or oligomerization are fused to CRY2-mCherry. In absence of the light the receptor-CRY2-mCherry does not oligomerize and thus remains inactive. Upon blue light exposure, the receptor-CRY2-mCherry oligomerizes and causes the signal cascade activation and translocation of nuclear effector; e.g., Opto-Egfr, Opto-Fgfr, CRY-Toll-mCherry[28,34,36,37].

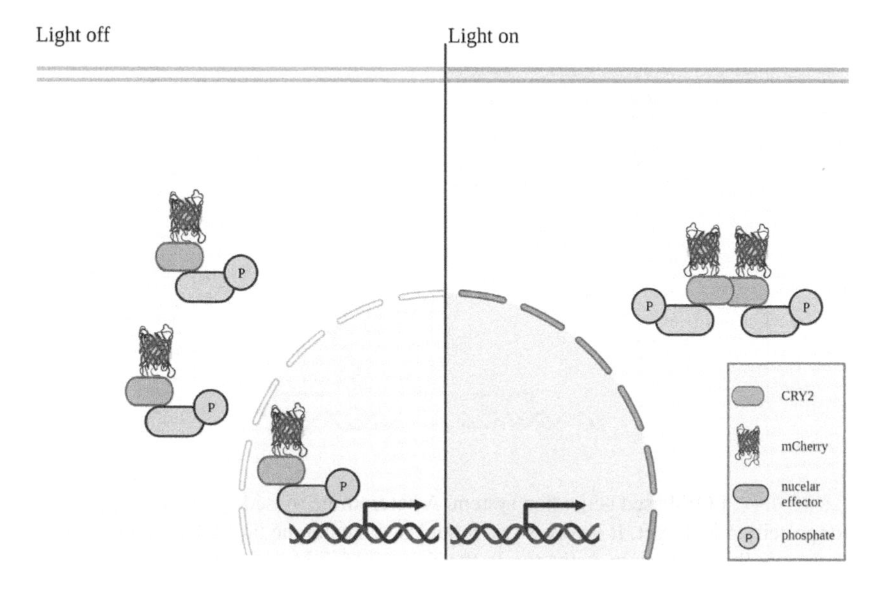

FIGURE 8.3 CRY2-based signal inhibition. CRY2-mCherry was fused to nuclear effectors of signalling pathways. In the absence of blue light, the nuclear effector can translocate to the nucleus and activate transcription of its target genes. Application of blue light will cause the nuclear effector to oligomerize inhibiting its nuclear translocation; e.g., β-catenin, ERK, Dorsal[28,34].

can also be applied directly to a transcription factor, using CRY2-based clustering to prevent nuclear localization (for example: Bicoid,[35]).

8.6 LOV-BASED OPTOGENETIC SYSTEM

8.6.1 Opto-SOS

LOV domain-based optogenetic system can either work in combination with other proteins as a heterodimer[13] or can act alone[38]. Opto-SOS is an important tool as it activates the MAPK cascade independently of RTK signal activation. SOS, a RAS activator, was combined with LOV to make this optogenetic tool using the iLID/SspB heterodimerization pair[39]. In this system the LOV protein from *Avena sativa* was modified by introducing mutations such that there was a considerable difference in its activity in dark versus light conditions. A membrane anchor was attached to iLID bearing a bacterial SsrA peptide that can bind to its partner SspB, which in turn is fused with SOS. In the absence of light, the SsrA is caged in the LOV protein conformation, but upon activation with blue light, the C terminus of LOV protein unwinds to expose the SsrA peptide and restore its interaction with SspB to recruit SOS toward the membrane, where it activates Ras/MAPK (ERK) cascade (Figure 8.4). This

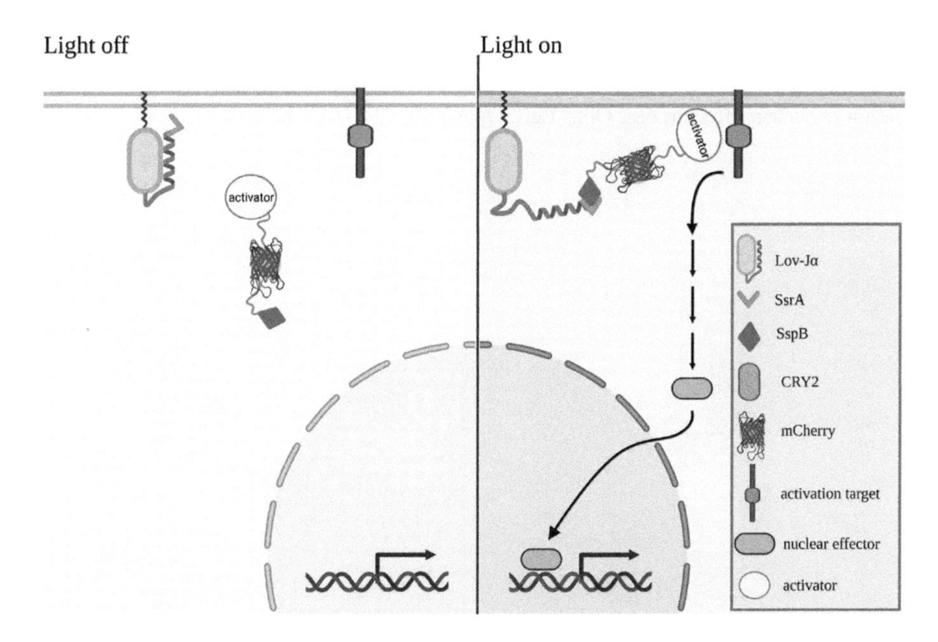

FIGURE 8.4 LOV-based activation system. A heterodimer-based system to recruit an activator protein to its target. It involves two modules, a membrane targeted Lov-Jα-SsrA and a cytoplasmic SspB-mCherry-Activator. In the absence of light, the two modules do not interact, as SsrA is hidden in Lov-Jα. With light, SsrA is exposed allowing SspB to interact with it, recruiting the activator protein to its target, activating it; e.g., Opto-SOS [13]. Other possible activator-target systems are GSK-3β or Dishevelled for canonical and non-canonical Wnt signaling and Myd88 for Toll signaling [42,43].

LOV-based Opto-SOS system is much improved in comparison to the previous Opto-SOS based on a phytochrome system[40,41].

8.7 OPTOGENTIC SYSTEMS BASED ON CAGING

8.7.1 Opto-YAP

The Hippo pathway is activated in a variety of ways[44]. The pathway culminates in the nuclear localization of YAP leading to transcriptional changes. To modulate the activity of Hippo signaling, Opto-YAP was developed. In the Opto-YAP construct the LOV protein is fused to the YAP protein and to a Nuclear Export Signal (NES). The construct is made in such a way that the Nuclear Localization Sequence (NLS) of YAP protein is in between the J-α helix and the LOV domain in the absence of blue light. After induction with blue light the NLS is exposed, and Opto-YAP can translocate to the nucleus to act on its target genes (Figure 8.5A). When blue light is turned off, the NLS is hidden and Opto-YAP is exported from the nucleus due to the NES[38].

8.7.2 DRONPA-Derived Photoswitchable Kinases

DRONPA is a photoswitchable green fluorescent protein that converts into a non-fluorescent form in the presence of 488 nm light and reverses back to the fluorescent form in presence of 405 nm light[45]. Though the original DRONPA protein is reported to exist as a monomer in solution the lattice structure shows it is also capable of adopting a tetramer state[46]. By introducing mutations, a photo-dissociable dimeric DRONPA (pdDRONPA) was generated, which is converted into dark monomeric pdDRONPA in the presence of 488 nm light and converts back into the bright dimeric pdDRONPA form in the presence of 405 nm light[47]. By attaching two pdDRONPA domains at a particular location in the kinase domain, photoswitchable kinases have been generated[15,47]. psMEK is an example of a photoswitchable kinase made by attaching two pdDRONPA domains to MEK, one at the N terminus and one in the flexible region of MEK (Figure 8.5B). pdDRONPA cages MEK's active site. Upon blue light exposure (488 nm), the active site of the enzyme is uncaged and thus can phosphorylate its target ERK. The benefit of using psMEK is that it acts directly on ERK rather than on upstream cascade components, which could have other effects. These approaches could be applied to other transcription factors and kinases.

8.8 OPTOGENETICS AND MICROSCOPY

As optogenetic tools have advanced and been constantly evolved, there has been a need for substantial improvements to the microscopy techniques used to detect their effects. To achieve desired results with optogenetic experiments, imaging parameters need to be well controlled. Parameters to be considered are the intensity and power of the light, speed of the microscope, viability of the sample, and imaging protocol modulation [48]. One microscopy technique, light-sheet microscopy, has become a preferred method to combine with optogenetics. Light-sheet microscopy

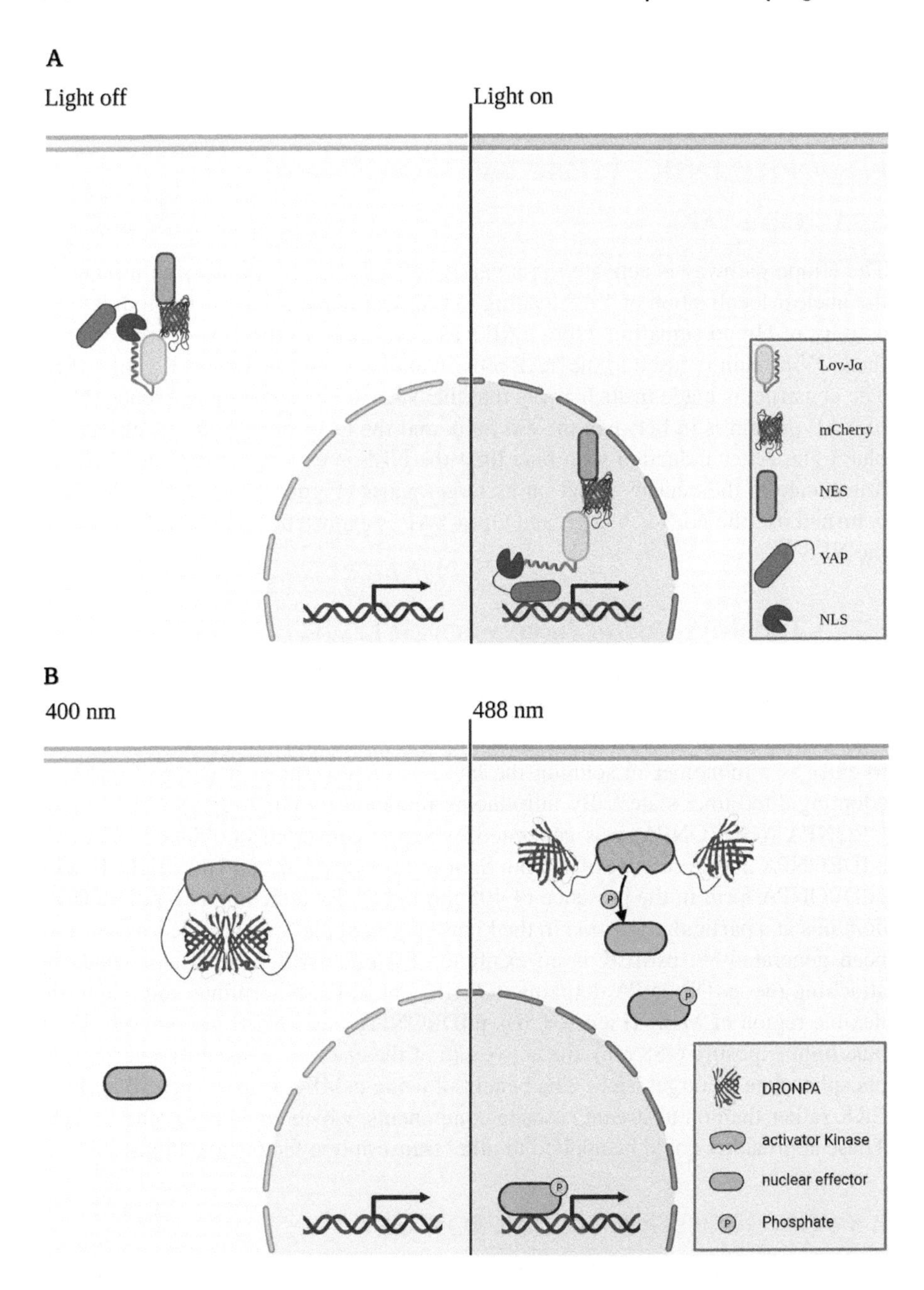

FIGURE 8.5 Caging-based optogenetic systems. (A) LOV-based caging. A transcription factor with an NLS is fused with Lov-Jα-mCherry-NES in such a way that NLS is caged and inaccessible in absence of blue light (488nm). As a consequence, the transcription factor is retained in the cytoplasm. Light causes a conformational change exposing the NLS leading to nuclear translocation; e.g., OptoYAP [38]. (B) DRONPA-based caging utilizes two DRONPA molecules fused to the protein of interest in such a way that the active site is blocked under UV light (400 nm) making enzyme inactive. Upon blue light activation, the active site becomes accessible and can act on its target protein; e.g., Opto-MEK [15].

has several advantages: compatibility with *in toto* imaging, lower laser power allowing the sample to be viable longer, and high speed of image acquisition. Light-sheet microscopy has been coupled with optogenetics to study the Wnt signal transduction pathway in *Drosophila* embryos, providing fine control in the excitation on/off regime [49]. It can be used to illuminate the sample partially to provide the internal control, thus enabling various ways to study the development of a tissue/organ [36]. In parallel, to investigate details at the cellular level, optogenetics is combined with confocal Laser Scanning Microscopy (LSM) [50,51]. Given that the long exposure to LSM is deleterious to cells and tissues, an alternative approach is to combine optogenetics with spinning disc confocal microscopy, which is less damaging. Further advances have been made in the microscopy to penetrate the deep tissue and counter light scattering by wavefront shaping [52]. Two-photon microscopy in combination to optogenetics is favourable to do extended deep tissues imaging [53,54].

8.9 DISCUSSION

This chapter has provided a glimpse of representative non-opsin-based optogenetic systems. There are a variety of approaches that we did not cover here, but growing interest in this field has led to many applications generating an enormous number of different optogenetic tools [51,55–57]. We have also provided some suggestions for extending the current approaches to other signaling pathways to make activators and inhibitors at different levels within the signal transduction cascades. For aggregation-based diseases, further models for Parkin or Amylin could be developed to observe the action of these proteins *in vivo*, perturb their activity in specific tissues, and to look for treatments that would ameliorate their detriments. As more approaches are used, and more laboratories become involved, the future of optogenetics in cell biology is indeed bright.

ACKNOWLEDGMENTS

We are thankful for the funding provided by AcRF grants IG19-SI102 and IG20-BG101to NST. PT and NST wrote the manuscript. We thank Prameet Kaur for useful discussions. The figures were made with Biorender.com.

REFERENCE

1. Deisseroth, K. Optogenetics. *Nat Methods* **8**, 26–29, doi:10.1038/nmeth.f.324 (2011).
2. Boyden, E., Zhang, F., Bamberg, E. Nagel, and G., Deisseroth, K. Millisecond-timescale, genetically targeted optical control of neural activity. *Nat Neurosci* **10**, 1263–1268 (2005).
3. Dema, A., van Haren, J. & Wittmann, T. Optogenetic EB1 inactivation shortens metaphase spindles by disrupting cortical force-producing interactions with astral microtubules. *Curr Biol* **32**, 1197–1205, e1194, doi:10.1016/j.cub.2022.01.017 (2022).
4. O'Neill, P. R., Kalyanaraman, V. & Gautam, N. Subcellular optogenetic activation of Cdc42 controls local and distal signaling to drive immune cell migration. *Mol Biol Cell* **27**, 1442–1450, doi:10.1091/mbc.E15-12-0832 (2016).
5. Liu, R. et al. Optogenetic control of RNA function and metabolism using engineered light-switchable RNA-binding proteins. *Nat Biotechnol*, doi:10.1038/s41587-021-01112-1 (2022).

6. Losi, A., Gardner, K. H. & Moglich, A. Blue-light receptors for optogenetics. *Chem Rev* **118**, 10659–10709, doi:10.1021/acs.chemrev.8b00163 (2018).
7. Tan, P., He, L., Huang, Y. & Zhou, Y. Optophysiology: Illuminating cell physiology with optogenetics. *Physiol Rev* **102**, 1263–1325, doi:10.1152/physrev.00021.2021 (2022).
8. Oh, T. J., Fan, H., Skeeters, S. S. & Zhang, K. Steering molecular activity with optogenetics: Recent advances and perspectives. *Adv Biol (Weinh)* **5**, e2000180, doi:10.1002/adbi.202000180 (2021).
9. Lim, C. H. et al. Application of optogenetic Amyloid-beta distinguishes between metabolic and physical damages in neurodegeneration. *Elife* **9**, doi:10.7554/eLife.52589 (2020).
10. Guglielmi, G., Barry, J. D., Huber, W. & De Renzis, S. An optogenetic method to modulate cell contractility during tissue morphogenesis. *Dev Cell* **35**, 646–660, doi:10.1016/j.devcel.2015.10.020 (2015).
11. Gil, A. A. et al. Optogenetic control of protein binding using light-switchable nanobodies. *Nat Commun* **11**, 4044, doi:10.1038/s41467-020-17836-8 (2020).
12. Humphreys, P. A. et al. Optogenetic control of the BMP signaling pathway. *ACS Synth Biol* **9**, 3067–3078, doi:10.1021/acssynbio.0c00315 (2020).
13. Johnson, H. E. et al. The spatiotemporal limits of developmental Erk signaling. *Dev Cell* **40**, 185–192, doi:10.1016/j.devcel.2016.12.002 (2017).
14. Wang, H. et al. LOVTRAP: An optogenetic system for photoinduced protein dissociation. *Nat Methods* **13**, 755–758, doi:10.1038/nmeth.3926 (2016).
15. Patel, A. L. et al. Optimizing photoswitchable MEK. *Proc Natl Acad Sci U S A* **116**, 25756–25763, doi:10.1073/pnas.1912320116 (2019).
16. Hardy, J. A. & Higgins, G. A. Alzheimer's disease: The amyloid cascade hypothesis. *Science* **256**, 184–185, doi:10.1126/science.1566067 (1992).
17. Cummings, J. Lessons learned from alzheimer disease: Clinical trials with negative outcomes. *Clin Transl Sci* **11**, 147–152, doi:10.1111/cts.12491 (2018).
18. Duff, K. et al. Increased amyloid-beta42(43) in brains of mice expressing mutant presenilin 1. *Nature* **383**, 710–713, doi:10.1038/383710a0 (1996).
19. Fong, S. et al. Energy crisis precedes global metabolic failure in a novel Caenorhabditis elegans Alzheimer Disease model. *Sci Rep* **6**, 33781, doi:10.1038/srep33781 (2016).
20. Teo, E. et al. Metabolic stress is a primary pathogenic event in transgenic Caenorhabditis elegans expressing pan-neuronal human amyloid beta. *Elife* **8**, doi:10.7554/eLife.50069 (2019).
21. LaFerla, F. M., Green, K. N. & Oddo, S. Intracellular amyloid-beta in Alzheimer's disease. *Nat Rev Neurosci* **8**, 499–509, doi:10.1038/nrn2168 (2007).
22. Ferreira, S. T. & Klein, W. L. The Abeta oligomer hypothesis for synapse failure and memory loss in Alzheimer's disease. *Neurobiol Learn Mem* **96**, 529–543, doi:10.1016/j.nlm.2011.08.003 (2011).
23. Kaur, P. et al. Use of optogenetic amyloid-beta to monitor protein aggregation in drosophila melanogaster, danio rerio and caenorhabditis elegans. *Bio Protoc* **10**, e3856, doi:10.21769/BioProtoc.3856 (2020).
24. Bugaj, L. J., Choksi, A. T., Mesuda, C. K., Kane, R. S. & Schaffer, D. V. Optogenetic protein clustering and signaling activation in mammalian cells. *Nat Methods* **10**, 249–252, doi:10.1038/nmeth.2360 (2013).
25. De Ferrari, G. V. et al. Activation of Wnt signaling rescues neurodegeneration and behavioral impairments induced by beta-amyloid fibrils. *Mol Psychiatry* **8**, 195–208, doi:10.1038/sj.mp.4001208 (2003).
26. Jin, N. et al. Sodium selenate activated Wnt/beta-catenin signaling and repressed amyloid-beta formation in a triple transgenic mouse model of Alzheimer's disease. *Exp Neurol* **297**, 36–49, doi:10.1016/j.expneurol.2017.07.006 (2017).

27. Parr, C., Mirzaei, N., Christian, M. & Sastre, M. Activation of the Wnt/beta-catenin pathway represses the transcription of the beta-amyloid precursor protein cleaving enzyme (BACE1) via binding of T-cell factor-4 to BACE1 promoter. *FASEB J* **29**, 623–635, doi:10.1096/fj.14–253211 (2015).
28. Kaur, P. et al. Wnt signaling rescues amyloid beta-induced gut stem cell loss. *Cells* **11**, doi:10.3390/cells11020281 (2022).
29. Cadigan, K. M. & Nusse, R. Wnt signaling: A common theme in animal development. *Genes Dev* **11**, 3286–3305, doi:10.1101/gad.11.24.3286 (1997).
30. Herrera-Perez, R. M., Cupo, C., Allan, C., Lin, A. & Kasza, K. E. Using optogenetics to link myosin patterns to contractile cell behaviors during convergent extension. *Biophys J* **120**, 4214–4229, doi:10.1016/j.bpj.2021.06.041 (2021).
31. Lammers, R., Van Obberghen, E., Ballotti, R., Schlessinger, J. & Ullrich, A. Transphosphorylation as a possible mechanism for insulin and epidermal growth factor receptor activation. *J Biol Chem* **265**, 16886–16890 (1990).
32. Krishnamurthy, V. V., Hwang, H., Fu, J., Yang, J. & Zhang, K. Optogenetic control of the canonical Wnt signaling pathway during xenopus laevis embryonic development. *J Mol Biol* **433**, 167050, doi:10.1016/j.jmb.2021.167050 (2021).
33. Lusk, J. B., Lam, V. Y. & Tolwinski, N. S. Epidermal growth factor pathway signaling in drosophila embryogenesis: Tools for understanding cancer. *Cancers (Basel)* **9**, doi:10.3390/cancers9020016 (2017).
34. Bunnag, N. et al. An optogenetic method to study signal transduction in intestinal stem cell homeostasis. *J Mol Biol* **432**, 3159–3176, doi:10.1016/j.jmb.2020.03.019 (2020).
35. Huang, A., Amourda, C., Zhang, S., Tolwinski, N. S. & Saunders, T. E. Decoding temporal interpretation of the morphogen Bicoid in the early Drosophila embryo. *Elife* **6**, doi:10.7554/eLife.26258 (2017).
36. Yadav, V., Tolwinski, N. & Saunders, T. E. Spatiotemporal sensitivity of mesoderm specification to FGFR signalling in the Drosophila embryo. *Sci Rep* **11**, 14091, doi:10.1038/s41598-021-93512-1 (2021).
37. Krishnamurthy, V. V. et al. A generalizable optogenetic strategy to regulate receptor tyrosine kinases during vertebrate embryonic development. *J Mol Biol* **432**, 3149–3158, doi:10.1016/j.jmb.2020.03.032 (2020).
38. Toh, P. J. Y. et al. Optogenetic control of YAP cellular localisation and function. *EMBO* Reports (2022)e54401, https://doi.org/10.15252/embr.202154401.
39. Guntas, G. et al. Engineering an improved light-induced dimer (iLID) for controlling the localization and activity of signaling proteins. *Proc Natl Acad Sci U S A* **112**, 112–117, doi:10.1073/pnas.1417910112 (2015).
40. Goglia, A. G., Wilson, M. Z., DiGiorno, D. B. & Toettcher, J. E. Optogenetic control of ras/erk signaling using the Phy-PIF system. *Methods Mol Biol* **1636**, 3–20, doi:10.1007/978-1-4939-7154-1_1 (2017).
41. Toettcher, J. E., Weiner, O. D. & Lim, W. A. Using optogenetics to interrogate the dynamic control of signal transmission by the Ras/Erk module. *Cell* **155**, 1422–1434, doi:10.1016/j.cell.2013.11.004 (2013).
42. Dunn, N. R. & Tolwinski, N. S. Ptk7 and Mcc, unfancied components in non-canonical Wnt signaling and cancer. *Cancers (Basel)* **8**, doi:10.3390/cancers8070068 (2016).
43. Nusslein-Volhard, C. The toll gene in drosophila pattern formation. *Trends Genet* **38**, 231–245, doi:10.1016/j.tig.2021.09.006 (2022).
44. Misra, J. R. & Irvine, K. D. The hippo signaling network and its biological functions. *Annu Rev Genet* **52**, 65–87, doi:10.1146/annurev-genet-120417–031621 (2018).
45. Habuchi, S. et al. Reversible single-molecule photoswitching in the GFP-like fluorescent protein Dronpa. *Proc Natl Acad Sci U S A* **102**, 9511–9516, doi:10.1073/pnas.0500489102 (2005).

46. Wilmann, P. G. et al. The 1.7 A crystal structure of Dronpa: A photoswitchable green fluorescent protein. *J Mol Biol* **364**, 213–224, doi:10.1016/j.jmb.2006.08.089 (2006).
47. Zhou, X. X., Fan, L. Z., Li, P., Shen, K. & Lin, M. Z. Optical control of cell signaling by single-chain photoswitchable kinases. *Science* **355**, 836–842, doi:10.1126/science.aah3605 (2017).
48. Huisken, J. & Stainier, D. Y. Selective plane illumination microscopy techniques in developmental biology. *Development* **136**, 1963–1975, doi:10.1242/dev.022426 (2009).
49. Kaur, P., Saunders, T. E. & Tolwinski, N. S. Coupling optogenetics and light-sheet microscopy, a method to study Wnt signaling during embryogenesis. *Sci Rep* **7**, 16636, doi:10.1038/s41598-017-16879-0 (2017).
50. Herrera-Perez, R. M. & Kasza, K. E. Manipulating the patterns of mechanical forces that shape multicellular tissues. *Physiology (Bethesda)* **34**, 381–391, doi:10.1152/physiol.00018.2019 (2019).
51. Hartmann, J., Krueger, D. & De Renzis, S. Using optogenetics to tackle systems-level questions of multicellular morphogenesis. *Curr Opin Cell Biol* **66**, 19–27, doi:10.1016/j.ceb.2020.04.004 (2020).
52. Yoon, J. et al. Optogenetic control of cell signaling pathway through scattering skull using wavefront shaping. *Sci Rep* **5**, 13289, doi:10.1038/srep13289 (2015).
53. Singh, A. P. et al. Optogenetic control of the bicoid morphogen reveals fast and slow modes of gap gene regulation. *Cell Rep* **38**, 110543, doi:10.1016/j.celrep.2022.110543 (2022).
54. Miller, D. R., Jarrett, J. W., Hassan, A. M. & Dunn, A. K. Deep tissue imaging with multiphoton fluorescence microscopy. *Curr Opin Biomed Eng* **4**, 32–39, doi:10.1016/j.cobme.2017.09.004 (2017).
55. Simpson, J. H. & Looger, L. L. Functional imaging and optogenetics in drosophila. *Genetics* **208**, 1291–1309, doi:10.1534/genetics.117.300228 (2018).
56. Johnson, H. E. & Toettcher, J. E. Illuminating developmental biology with cellular optogenetics. *Curr Opin Biotechnol* **52**, 42–48, doi:10.1016/j.copbio.2018.02.003 (2018).
57. Goglia, A. G. & Toettcher, J. E. A bright future: Optogenetics to dissect the spatiotemporal control of cell behavior. *Curr Opin Chem Biol* **48**, 106–113, doi:10.1016/j.cbpa.2018.11.010 (2019).

9 Opsin-free Optogenetics: Brain and Beyond

Jongryul Hong, Yeonji Jeong, and Won Do Heo

CONTENTS

9.1 Introduction ... 123
9.1.1 Photoactivatable Flp Recombinase (PA-Flp) ... 124
9.1.2 monSTIM1 ... 124
9.1.3 Opto-cytTrkB(E281A) ... 124
9.1.4 Opto FAS ... 125
9.2 Materials and Methods ... 125
9.2.1 Plasmids ... 125
9.2.2 Viral Vectors Preparation ... 126
9.2.3 LV ... 126
9.2.4 AAV ... 127
9.2.5 Virus Injection and Fiber Optic Cannula Implantation into the Mouse Brain ... 129
9.2.6 Light Delivery ... 130
9.3 Conclusion ... 132
Acknowledgments ... 132
References ... 133

9.1 INTRODUCTION

The ability to modulate the activity of signaling molecules or enzymes is key to investigating their contributions to molecular, cellular, and behavioral changes. Overexpression of target proteins or the use of agonists or ligands that modulate their activity have been widely employed strategies for regulating specific signaling pathways or gene expression. Although these methods have provided approaches for addressing diverse questions in the biological field, they have critical limitations in that they often fail to deliver spatial and/or temporal precision *in vivo*.

Moving beyond the cellular level to *in vivo* studies necessitates the tools which become versatile and controllable without producing adverse effects. Opsin-free optogenetic tools have notable advantages given that their activities are controlled by light in a highly spatiotemporal manner (Tischer and Weiner, 2014). Especially in neurobiology, optogenetic tools are now widely used for elucidating neural mechanisms, circuits, behaviors, as well as the function of non-neuronal cells (Oh et al., 2021). The tools have opened new avenues to probe sophisticated *in vivo* processes that are otherwise experimentally challenging or intractable.

DOI: 10.1201/b22823-9

Over the past five years we have developed a set of innovative light-sensitive recombinases (Jung et al., 2019), signaling molecules and receptors which showed blue light-induced activities or signal transductions from the viral vector-mediated expression in the mouse brain (Hong and Heo, 2020; Kim et al., 2020a). Here, we will discuss the detailed methods and protocols with brief introductions to *in vivo* studies.

9.1.1 Photoactivatable Flp Recombinase (PA-Flp)

Jung et al. developed a non-invasive optogenetic Flp recombinase (PA-Flp) for manipulating gene expression in the mouse brain (Jung et al., 2019). The PA-Flp was constructed as follows: FlpO was split at the position of residues 27–28 in Flp, and the two split FlpO fragments were fused to nMagHigh1 and pMagHigh1, respectively (Kawano et al., 2015). The tool displays remarkable sensitivity toward blue light, making it suitable to remotely control recombinase activity in deep brain. Furthermore, Jung et al. extended its application to the Cre-loxP system with leak-free Flp-dependent Cre driver (LF-FdCd), thus demonstrating that the mice with Cre-mediated $Ca_v3.1$ knock-down in the medial septum showed an increase in object exploration behavior.

9.1.2 monSTIM1

Kim et al. developed an optogenetic toolkit, monSTIM1 (Kim et al., 2020b). The monSTIM1 consists of cytosolic domains of STIM1 (stromal interaction molecule 1) protein as the effector module and blue light-responsive plant photosensory module, AtCRY2 (*Arabidopsis thaliana* cryptochrome 2), which is modified for higher photosensitivity. As previously reported, photoreceptor AtCRY2 (*Arabidopsis thaliana* Cryptochrome2) under 400–500 nm light undergoes oligomerization. CRY2 with blue light allows the cytosolic domain of STIM1 to be oligomerized and translocated toward the plasma membrane to engage and gate endogenous ORAI channels and trigger Ca^{2+} influx. It has ultra-light sensitivity that can be activated *in vivo* through non-invasive light illumination and directly induce intracellular Ca^{2+} signals to study their impact on behaviors in living mice.

9.1.3 Opto-cytTrkB(E281A)

Hong et al. improved the optogenetic TrkB receptor (Opto-cytTrkB(E281A)) to be more compatible with *in vivo* applications, which enabled light-induced transient/sustainable signaling in the mouse brain (Hong and Heo, 2020). The Opto-cytTrkB(E281A) system was built based on the original Opto TrkB (Chang et al.) with two major modifications: substitution of the extracellular domain of TrkB with a membrane-tethering sequence (Lyn) and the point mutation of Cry2PHR(E281A). With this tool, Hong et al. further demonstrated the use of pulsed illumination for precise signal activation and axon-specific TrkB signaling activation, which cannot be accomplished by direct infusion of ligands.

9.1.4 Opto FAS

Kim et al. developed an optogenetic FAS (CD95) receptor to investigate dynamic Fas signaling in the mouse hippocampal dentate gyrus (DG) (Kim et al., 2020a). The tool, consisting of the Lyn-cytosolic Fas domain-Cry2PHR-EGFP, exerted cell death signaling via blue light illumination in cultured HeLa cells. However, distinct signaling mechanisms were observed in hippocampal DG. Kim et al. demonstrated that Fas signaling could be photo-manipulated to promote adult neurogenesis and working memory enhancement in living rodents.

9.2 MATERIALS AND METHODS

9.2.1 Plasmids

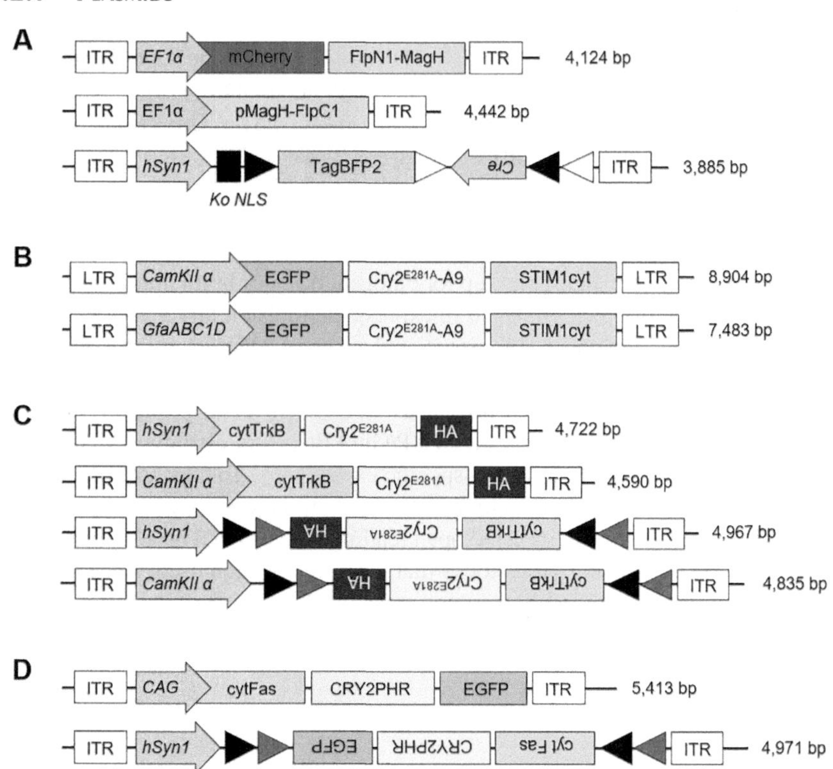

FIGURE 9.1 DNA construct of each optogenetic toolkits. (A) PA-Flp constructs. pAAV-EF1α::mCherry-FlpN-nMagHigh1, pAAV-EF1α::pMagHigh-FlpC, pAAV-hSyn1::LF-FdCd (hSyn1::KozNLS-fDIO-(tagBFP2)-Cre). (B) monSTIM1 constructs. pLenti-CaMKIIα-EGFP-monSTIM1 for excitatory neuron expression, pLenti-GfaABC1D-EGFP-monSTIM1 for astrocyte expression. (C) Opto-cytTrkB(E281A) constructs. pAAV- hSyn1::Opto-cytTrkB (E281A)-HA and pAAV-CamkIIα::Opto-cytTrkB(E281A)-HA for neuron expression. pAAV-hSyn1::DIO-Opto-cytTrkB(E281A)-HA, pAAV-CamkIIα::DIO-Opto-cytTrkB(E281A)-HA for Cre dependent expression in neurons. (D) Opto FAS constucts. pAAV-CAG::OptoFAS for universal expression and pAAV-hSyn1::DIO-OptoFAS for neuron expression.

Main plasmids

- PA-Flp: pAAV-*EF1*α::mCherry-FlpN-nMagHigh1, pAAV-*EF1*α::pMagHigh-FlpC, pAAV-*hSyn1*::LF-FdCd(hSyn1::KozNLS-fDIO-(tagBFP2)-Cre) (Figure 9.1A).
- monSTIM1: pLenti-*CaMKII*α-EGFP-monSTIM1, pLenti-*GfaABC1D*-EGFP-monSTIM1 (Figure 9.1B).
- Opto-cytTrkB(E281A): pAAV-CamkIIα::Opto-cytTrkB(E281A)-HA, pAAV-hSyn1::Opto-cytTrkB(E281A)-HA, pAAV-CamkIIα::DIO-Opto-cytTrkB (E281A)-HA, pAAV-hSyn1::DIO-Opto-cytTrkB(E281A)-HA (Figure 9.1C).
- Opto-FAS: pAAV-*hSyn1*::DIO-OptoFAS, pAAV-*CAG*::OptoFAS (Figure 9.1D).

Auxiliary plasmids

FlpO (Addgene #55634), pAAV-*EF1*α::fDIO-EYFP (Addgene #55641), pAAV-*EF1*α::DIO-EGFP (Addgene #37084), pAAV-*hSyn1*-DIO-EGFP (Addgene #50457), R-GECO1 (Addgene #32444), pAAV-*hSyn1*-Cre (Hong et al), AAV-*Nestin*::Cre (Kim et al).

9.2.2 Viral Vectors Preparation

Gene delivery is an indispensable step for optogenetics. For brain optogenetics, viral vector-mediated delivery is preferred to express the tools on target regions and cell types. Lentivirus (LVs) and adeno-associated virus (AAVs) are mostly adopted to deliver the optogenetic tools into the mouse brain. To express the tools in the neurons, promoters of Camk*II*α and hSyn1, as well as universal promoter EF1α, are generally used. Other promoters (e.g., GFAP/GfaABC1D for astrocyte; MAG for oligodendrocyte) may provide cell type-specific expressions. To take advantage of cell type-specific expression, we recommend the tools to be expressed under the common promoter with a double inverted open reading frame (DIO) on Cre lines. We do not recommend re-freezing viral stocks that had already been used, since repetitive freeze-thawing procedures tend to compromise the stability. In this section, we will present protocols for producing AAVs and LVs, along with their strengths and weaknesses for *in vivo* usage.

9.2.3 LV

The large packaging capacity (up to 9 kb) of LV enables broader applications that require extensive optogenetic tools compared to AAVs. The gene delivery with lentivirus is suitable when the tools have large actuators or photoreceptors like Cry2PHR. Moreover, the lentivirus can infect and integrate into the host genome in dividing and non-dividing cells (Ciuffi, 2008). Thus, the lentivirus can infect neurons, primarily regarded as non-dividing cells in the brain. But the tropism for infection has not yet been well expanded. However, we observed that LVs have a rather narrow expression area coverage when injected into the brain, and it takes over 4 weeks to achieve an appropriate expression level (Kyung et al., 2015; Kim et al., 2020b)

- LVs packaging, purification and titration
  1. For LV production, culture HEK293T cells (ATCC CRL-3216) in DMEM (Gibco 11995073) with 10% FBS (Gibco 12483020), 1% penicillin/streptomycin (Gibco 15140122).
  2. Plate the HEK293T cells on 5 × 150 Φ culture dishes and grow up the HEK29T cells until 70~80% confluency.
  3. Transfect the HEK293T cells with DNA (20 μg lentiviral vector/plate, 15 μg delta8.9 (Labome PVT2323)/plate,10 μg VSV-G (Addgene #8454)/plate) with 140 μg of 4 μg/ml polyethyleneimine (PEI, Sigma 919012) in 5 ml DPBS (Gibco 14190144).
  4. Change fresh culture media after 6 hr.
  5. 48 hr after transfection, collect virus-containing media in 50 ml tube and centrifugate at 600 g for 5 min to remove debris.
  6. Transfer supernatant to syringe filters with a 0.45 μm pore size to 40 ml Ultraclear Bechman tube (Beckman #344058) and seal the top with parafilm (Bemis Film).
  7. Centrifugate (Rotor 32Ti) at 77,000 g for 90 min at 4°C.
  8. After discarding the supernatant and completely removing it, invert the tube on Kimwipe for 1 min, and gently dissolve the virus in DPBS overnight utilizing a rotator.
  9. Perform the titration of LVs with Lenti-X™ qRT-PCR Titration Kit (TaKaRa 631235).
  10. Aliquot the virus and store it at -80°C and used within 12 months.

9.2.4 AAV

AAV has a restricted packaging size limit (~ 4.9 kb) (Grieger and Samulski, 2005; Wu et al., 2010), but is widely chosen for *in vivo* research due to several advantages. It takes less time for a high expression compared to lentivirus. Gene delivery using AAV provides a wide coverage of expression when injected into the brain. The expression area could be further regulated by adjusting the injection volume or titer of the virus. In addition, diverse capsid variants allow AAVs to infect multiple cell types in different brain regions with varying expression levels and tropisms, making AAVs more suitable for specific gene delivery into the brain and other organs (Aschauer et al., 2013; Tervo et al., 2016; Zingg et al., 2017; Challis et al., 2019; Goertsen et al., 2021). Capitalizing on these desirable attributes, we often express our optogenetic tools with AAVs in rodent brains.

- AAVs packaging & purification
  1. Culture the HEK293T cells (ATCC CRL-3216) under the DMEM (Gibco 11995073) with 10% FBS (Gibco 12483020) without any antibiotics.
  2. Scale up the HEK293T cells until 70–80% of confluency of 10 × 150 Φ culture dishes for each pAAV vector (10 dishes for 1 viral vector) (Figure 9.2A).
  3. Transfect the HEK293T cells with DNAs (Target vector 41.5 μg, Capsid 166 μg, Helper plasmid 83 μg) with 120 μl of 4 mg/ml PEI (Sigma

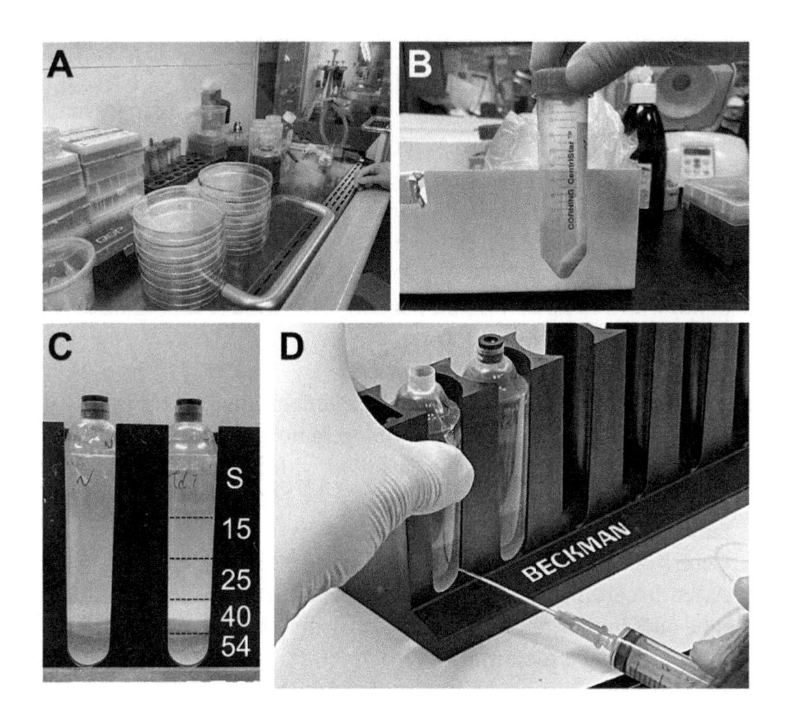

FIGURE 9.2 Preparation of AAVs.

Steps of AAVs production. (A) Cultured HEK293T cells in 150-mm culture dishes for AAVs. (B) Crude AAVs in lysis buffer after the freeze-thaw with centrifugation steps. (C) Iodixanol buffer layers in an ultracentrifuge tube. (S: sample, 15: 15% of iodixanol, 25: 25% of iodixanol, 40: 40% of iodixanol, 54: 54% of iodixanol. (D) Image showing the collection of the 40% iodixanol layer that contains AAVs.

919012) in 15 ml of DPBS (Gibco 14190144). Change the culture media 6 hr after the transfection.

4. 48 hr after the transfection, collect the media and store at 4°C until the second prep. Culture the remaining cells with fresh culture media.
5. 120 hr after transfection, collect the media and cells in step #4. Centrifugate at 3,300 g for 20 min, then mix the supernatant with 40% of NaCl-PEG solution to make 8% PEG-media solution (NaCl-PEG: 2.5 N NaCl with 40% PEG 8000 (Sigma P2139)). React for 2 hr at 4°C, then centrifugate again at 2,500 g for 30 min. Discard the solution.
6. Mix the PEG-precipitated debris with centrifuged cells with 9 ml of lysis buffer (50 mM Tris-Cl, 150mM NaCl, 2mM $MgCl_2$, pH at 8). Add 3 µl of nuclease (Sigma E1014) with 482 ml of 10% of sodium deoxycholate (Sigma D6750) to lysis the cells. React for 45 min at 37°C.
7. Freeze-thaw the mixture for further lysis and aggregate the debris at –80°C to 37°C three times. Centrifugate at 12,000 g for 30 min. Collect the supernatant (Figure 9.2B).

8. Prepare the ultracentrifuge tube and iodixanol gradient buffers (Tube: Beckman 361625. Iodixanol: Progen 1114542). Stack the tube in order of sample (9 ml), 15% (6 ml), 25% (5 ml), 40% (5 ml) and 54% (5 ml) of iodixanol buffers from top to bottom. (15% of iodixanol buffer: 1 M NaCl and 15% of iodixanol in PBS-MK (PBS-$MgCl_2$/KCl.) 25% of iodixanol buffer: 25% of iodixanol with 0.1% of phenol-Red (Sigma P3532) in PBS-MK. 40% of iodixanol buffer: 40% of iodixanol in PBS-MK. 54% of iodixanol buffer: 54% of iodixanol with phenol-Red in PBS-MK. PBS-MK: 10 mM $MgCl_2$, 25 mM KCl in 10XPBS (working at 1X). Centrifugate at 350,000 g for 1 hr using a 70Ti rotor for ultracentrifuge (Beckman Optima XE-100) (Figure 9.2C).
9. Collect 40% of iodixanol stack and filtrate through 0.2 μm syringe filter (Sartorius 16534K). Wash with final buffer (e.g., PBS) over 4 times in filter concentrator (Millipore UFC910024) (Figure 9.2D).
10. Perform the titration of AAVs with AAVpro Titration Kit (for Real time PCR, TaKaRa 6223).
11. Aliquot and store at –80°C and use within 12 months.

9.2.5 Virus Injection and Fiber Optic Cannula Implantation into the Mouse Brain

These methods and protocols are considered for cranial injection by following automated stereotaxic procedures (Cetin et al., 2006). The following precautions should be taken during the surgery. First, avoid second surgery after virus injection. We strongly recommend performing one-step surgery with fluid injection and device implantation to minimize undesired activation of the tools via light source during a subsequent surgery. Second, pay careful attention to suturing the incision. Due to the natural characteristics of mice, the skin would be torn if your suturing is poor, which could result in unexpected activation of the tools via a room light source. Lastly, the use of an appropriate viral titer is important. Too high or low viral titer could lead to sub-optimal expression of the optogenetic tools in the mouse brain. For example, a high titer of lentivirus of monSTIM1 is required to achieve appropriate activating conditions. However, low titers of AAVs seem to work best for other optogenetic receptors (Opto-cytTrkB, Opto Fas). Also, precise titers of each AAVs encoding N-Flp and C-Flp are required to generate PA-Flp with maximal functionality. The following protocol describes the procedures of cranial fluid microinjection and optic fiber implantation.

1. Animal experiments and treatments should follow the guidelines of the institutional animal usage and treatment ethics and policy.
2. Prepare the subjects (hereafter, mice). Anesthetize the mice with Avertin (0.2 ml/10 g). (Avertin: 0.5 ml of stock (1,600 mg/ml) solution with 39.5 ml of saline. Stock solution: 2,2,2-tribromoethanol (Sigma T48402) dissolved in 2-methyl-2-butanol (Sigma 152463).
3. Incise the fur coat and drill the target skull area.

4. Stab the glass needle tip into the target region (1 mm/min), then inject the fluids at a rate of 0.5–0.7 μl/min. Wait for 10 min to prevent the fluids from back flowing. Withdraw the glass needle.
5. Implant the fiber optic cannulas (Doriclenses) typically 50–100 μm above the target regions. Nonetheless, it depends on the area and light power to be delivered to the brain. Fix the cannulas on the skull using dental cement. (Dental cement: catalyst superbond, L-type radiopaque polymer superbond, monomer superbond.) This step is not needed for transcranial illumination experiments.
6. Suture the incision and return the mice to their home cages for recovery.

9.2.6 Light Delivery

Delivery of light sources is mandatory for effective activation of optogenetic tools. To enable efficient activation of optogenetic tools in tissues, one should take into account the types of light delivery as well as the respective illuminating conditions. A fiber-optic cannula is often used to deliver visible light into the mouse brain. This method enables facile control of illuminating conditions in the mouse brain, which allows flexible adjustment of the light intensity, frequency, and illumination areas with fibers of varying diameters. However, the durability and reusability of optic equipment and parts remain troublesome because mice often gnaw the optic patch cord during long-term stimulation. Transcranial illumination, as discussed subsequently, might offer a solution to solve this issue, but at the cost of sacrificing high spatial resolution. In this section, we will discuss the methods of light delivery for the non-opsin-based optogenetic tools.

- Fiber optic cannula
  1. Before the optogenetic activation, handling the mice is strongly recommended in the room where the animals will be placed during the experiments. Let the mice freely move around the hands of a researcher for 5 min. Repeat the handling 3 days before the optogenetic experiments.
  2. Set the pulse types by adjusting the light power. The types of pulse and light intensity may be variable depending on experimental needs and equipment setups. Our group has set the pulse generator (Keysight technologies 33511B) with ~ 1–5 mW/mm^2 of 473 nm DPSS Blue Laser (CNI MBL-III-473) (Figure 9.3A to C). We found that the duration of light illumination or types of the pulse are important for setting suitable conditions for the non-opsin-based optogenetics tools rather than light intensity itself. The signaling or enzyme activity actuated by the optogenetic tools has a switch-like property in the mouse brain (i.e., partial activation of the tools would not be possible), postulating that the active-inactive time of the tools relies on the kinetics of engineered photoreceptors. Also, consider the effect of scattered light from the optic ferrule tip to the target area. Overall, we recommend that the light stimulating conditions be confirmed before the experiments.

3. Connect the tip of a cannula with an optic patch cord. Fix the mice heads with a gentle grabbing of the body (Figure 9.3D). We recommend that this step be finished as quickly as possible to minimize distress and pain. Let the mice move around the home cages or under appropriate experimental conditions (Figure 9.3E).
4. After finishing the experiment, detach the optic patch cord from the cannula.

- Transcranial illumination

1. We recommend handling the mice in the room where the light illumination will be performed. The customized transcranial cage is equipped with a cage lid that emits blue light from LED sources (Figure 9.4A and B). Also, mini-fans facilitate the cooling of LED to avoid heating effects (Figure 9.4C).
2. Set a proper light power. We set the power around 1 mW/mm^2 on the cage floor, which is adequate to switch on the optogenetic tools in the mouse brain. There is no need to incise head skin or barber the fur coat. Also, we demonstrated that room light (300–500 lux) is not sufficient to activate the tools, so the basal activity via the room light is negligible. We have not found the defects of transcranial illumination under the condition of

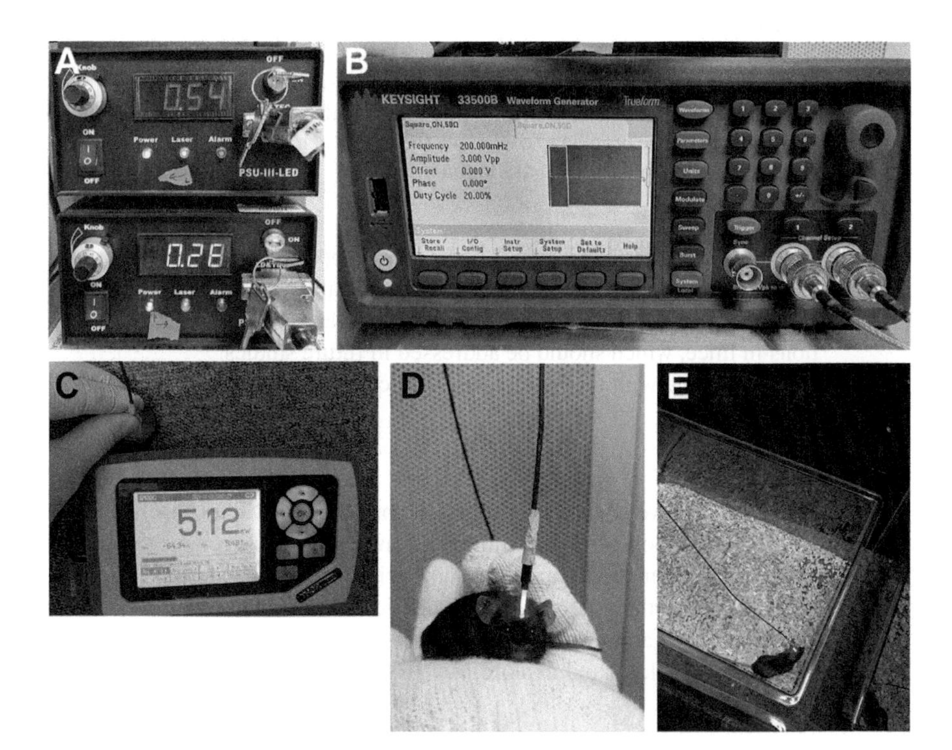

FIGURE 9.3 Light delivery with fiber optic cannula. Equipment for photo-stimulation setting and light delivery into the mouse brain. (A) 473 nm laser. (B) Pulse generator. Set with 1 s per 5 s stimulation. (C) Measuring the light intensity. (D) A mouse connected with an optic patch cord. (E) A freely moving mouse with blue light stimulation.

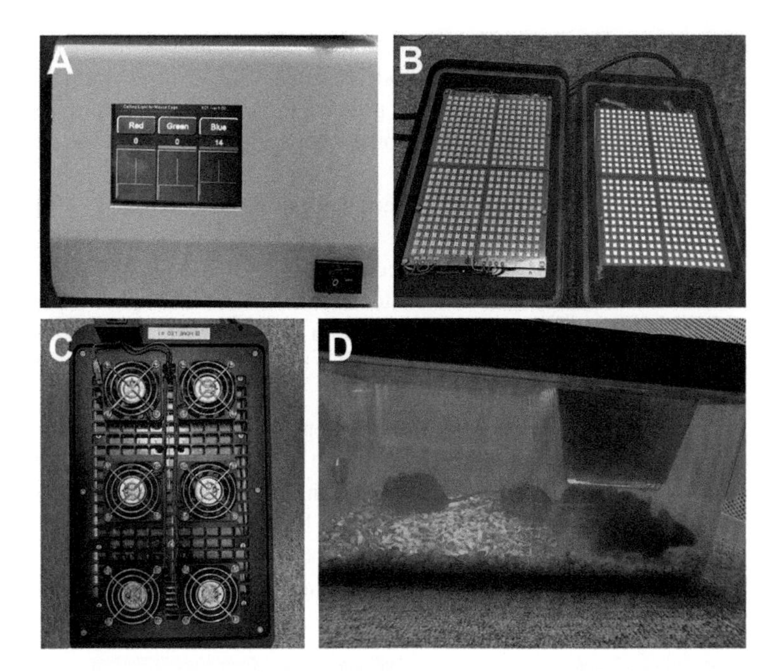

FIGURE 9.4 Light delivery with transcranial illumination. Equipment of transcranial optogenetics and light delivery. (A) Customized transcranial modulator. (B) The customized LED cage lid. (C) Mini-fans on the cage lid to cool down the heat generated from the LED. (D) The freely moving mice under the transcranial illuminating cage.

1 mW/mm^2, observing normal behaviors compared to control groups. However, be aware that stimulation with substantial power of light for a long period may damage visual perception or cause psychological conditions in mice, which should be addressed in future studies.

3. Cover the cage with the lid and let the mice freely move (Figure 9.4D).

9.3 CONCLUSION

Optogenetic methods have been widely used to control the activation/inactivation of cellular signaling, morphogenesis, and animal behaviors. The optogenetic tools allow researchers to easily manipulate target proteins without adding external ligands or substrates and to control their activities with superior spatio-temporal precision both *in vitro* and *in vivo*. We anticipate that non-opsin-based optogenetics will pave the way to discover new biology and explore new frontiers that are inaccessible due to the lack of suitable tools.

ACKNOWLEDGMENTS

The introduced work was supported by Institute for Basic Science. The introduced work is based on a research conducted as part of the KAIST-funded Global Singularity Research Program for 2021. The introduced work was supported by the

National Research Foundation of Korea (NRF-2020R1A2C3014742). The introduced work was supported by KAIST Institute for the BioCentury, Republic of Korea. We thank Dr. Kai Zhang for giving great pioneering works in the optogenetics fields. We thank Dr. Yubin Zhou for sharing the broad leadership in the optogenetics fields.

REFERENCES

Aschauer, D. F., et al. (2013). "Analysis of transduction efficiency, tropism and axonal transport of AAV serotypes 1, 2, 5, 6, 8 and 9 in the mouse brain." *PLoS One* **8**(9): e76310.

Cetin, A., et al. (2006). "Stereotaxic gene delivery in the rodent brain." *Nat Protoc* **1**(6): 3166–3173.

Challis, R. C., et al. (2019). "Systemic AAV vectors for widespread and targeted gene delivery in rodents." *Nat Protoc* **14**(2): 379–414.

Ciuffi, A. (2008). "Mechanisms governing lentivirus integration site selection." *Curr Gene Ther* **8**(6): 419–429.

Goertsen, D., et al. (2021). "AAV capsid variants with brain-wide transgene expression and decreased liver targeting after intravenous delivery in mouse and marmoset." *Nat Neurosci* **25**(1):106–115.

Grieger, J. C. and R. J. Samulski (2005). "Packaging capacity of adeno-associated virus serotypes: impact of larger genomes on infectivity and postentry steps." *J Virol* **79**(15): 9933–9944.

Hong, J. and W. D. Heo (2020). "Optogenetic modulation of TrkB signaling in the mouse brain." *J Mol Biol* **432**(4): 815–827.

Jung, H., et al. (2019). "Noninvasive optical activation of Flp recombinase for genetic manipulation in deep mouse brain regions." *Nat Commun* **10**(1): 314.

Kawano, F., et al. (2015). "Engineered pairs of distinct photoswitches for optogenetic control of cellular proteins." *Nat Commun* **6**: 6256.

Kim, S., et al. (2020a). "Dynamic Fas signaling network regulates neural stem cell proliferation and memory enhancement." *Sci Adv* **6**(17): eaaz9691.

Kim, S., et al. (2020b). "Non-invasive optical control of endogenous Ca(2+) channels in awake mice." *Nat Commun* **11**(1): 210.

Kyung, T., et al. (2015). "Optogenetic control of endogenous Ca(2+) channels in vivo." *Nat Biotechnol* **33**(10): 1092–1096.

Oh, T. J., et al. (2021). "Steering molecular activity with optogenetics: Recent advances and perspectives." *Adv Biol* (Weinh) **5**(5): e2000180.

Tervo, D. G., et al. (2016). "A designer AAV variant permits efficient retrograde access to projection neurons." *Neuron* **92**(2): 372–382.

Tischer, D. and O. D. Weiner (2014). "Illuminating cell signalling with optogenetic tools." *Nat Rev Mol Cell Biol* **15**(8): 551–558.

Wu, Z., et al. (2010). "Effect of genome size on AAV vector packaging." *Mol Ther* **18**(1): 80–86.

Zingg, B., et al. (2017). "AAV-mediated anterograde transsynaptic tagging: Mapping corticocollicular input-defined neural pathways for defense behaviors." *Neuron* **93**(1): 33–47.

10 Constructing a Far-Red Light-Induced Split-Cre Recombinase System for Controllable Genome Engineering

Meiyan Wang, Jiali Wu, and Haifeng Ye

CONTENTS

10.1 Introduction 136
10.2 Construction of an FRL-Induced Split Cre-loxP System 137
10.2.1 Optimization of the Different FRL-Responsive Promoters P_{FRLx} 137
10.2.1.1 Construction of Three Different FRL-Responsive Chimeric Promoters 137
10.2.1.2 Step-by-Step Showcase Protocol 138
10.2.2 Optimization of Different Linkers Between CreN59 and Coh2 as well as Between CreC60 and DocS 140
10.2.3 Optimization of the CreN59/CreC60 Catalytic Activity for Different *loxP* Mutants 140
10.3 DNA Recombination Performance of the FISC System in Mammalian Cells 140
10.3.1 Step-by-Step Showcase Protocol 140
10.3.1.1 Exposure-Time and Illumination-Intensity-Dependent FISC System Activity 140
10.3.1.2 Setup for Pattern Illumination 142
10.3.1.3 Spatial Control of FRL-Dependent DNA Recombination Mediated by the FISC System 142
10.4 Optogenetic DNA Recombination in tdTomato Transgenic Mice 143
10.4.1 *In vivo* DNA Recombination with the FISC System Using Hydrodynamic Injection Delivery 143
10.4.2 *In vivo* DNA Recombination with the FISC System Using AAV Delivery 143
10.4.3 Step-by-Step Showcase Protocol 145

DOI: 10.1201/b22823-10

10.4.3.1 Preparation of AAV-Mediated FISC System 145
10.4.3.2 FISC-Mediated DNA Recombination in Mice 146
10.4 Perspectives 147
Acknowledgments... 148
References... 148

10.1 INTRODUCTION

The versatile Cre-*loxP* recombination system has been widely used for genetic manipulation of mammalian cell lines and transgenic animals for a number of purposes such as cell fate mapping[1,2] and disease treatment[3,4] due to its simplicity and efficiency. Cre recombinase, a site-specific recombinase of the integrase family, catalyzes homologous DNA recombination between pairs of 34-bp sequences called *loxP* sites[5] and serves to induce or silence gene expression for conditional transgenesis or conditional knock-outs[6]. However, limitations exist in tight control over Cre activity and thus it can lead to early embryonic death, chromosomal rearrangements, and disturbances of cell physiology[7].

Over the past few decades, a variety of chemically inducible Cre-*loxP* systems such as tetracycline[8,9], tamoxifen[10,11] and rapamycin[12] controlled DNA recombination systems have been developed to achieve controllable genome engineering *in vivo*[13]. However, these systems exhibit unwanted side effects such as cytotoxicity of these chemicals, leakiness, and off-target DNA recombination. In this regard, innovative control strategies are needed to enable the spatiotemporal manipulation of Cre activity. Compared to chemicals, light is an orthogonal stimulus and highly adjustable. By controlling the time, intensity, and frequency of the light irradiation, it can achieve dynamical induction of gene expression with high spatiotemporal resolution. To achieve traceless and remote control of DNA recombination, several light inducible Cre-*loxP* systems have been developed mainly based on UV[14–16] and blue light[17–19]. However, the performance of these systems is unsatisfactory with its poor penetrative capacity through turbid human tissues and phototoxic effects on cells, which restricts these systems for further translational research and clinical applications[20].

To overcome this limitation, more advanced optogenetic inducible Cre-*loxP* systems need to be further explored. Light emitting in the longer wavelength (700–900 nm) can penetrate more deeply-buried living tissues and organs *in vivo*[21]. Therefore, we explored the generation of a far-red light (FRL) induced Split Cre-*loxP* (FISC) system, which is built by taking advantage of a previously reported split Cre recombinase and our validated far-red light-inducible transgene expression system[22]. In the FISC system, Cre recombinase is split into two non-active fragments, with the N-terminal Cre fragment (CreN (1–59)) fused to a Coh2 domain driven by a constitutive promoter P_{hCMV}. The C-terminal Cre fragment (CreC (60–343)) is fused to a DocS domain driven by the FRL-inducible promoter P_{FRLx}. Upon illumination with FRL, a functional Cre recombinase can be reassembled by taking advantage of the interaction between Coh2 and DocS[23]. This system exhibits low background leakage, non-toxic side effects, and high recombinant efficiency with spatiotemporal control. It allows for efficient DNA recombination *in vitro* and in internal organs of BALB/c wild-type mice and $Gt(ROSA)26Sor^{tm14(CAG\text{-}tdTomato)Hze}$ mice. Moreover, the FISC system can be delivered into mouse organs via adeno-associated virus (AAV) vectors, followed by successful induction of DNA recombination under FRL light-emitting diode (LED) light illumination. In this chapter, we present an overview of the procedures of constructing a far-red light-induced split Cre-*loxP* system for precise control of genome engineering in a spatiotemporal fashion with the capacity for deep penetration. This system allows for a previously unattainable level of non-invasive control that should facilitate studies and therapies focused on a broad range of biological processes.

10.2 CONSTRUCTION OF AN FRL-INDUCED SPLIT CRE-LOXP SYSTEM

To create an improved photoactivatable Cre-*loxP* system with low phototoxicity and deep tissue penetrative capacity, we have developed an FRL-induced Cre-*loxP* system (FISC system) based on the split-Cre recombinase and our previous bacteriophytochrome-based optogenetic system[22] (Figure 10.1). In this system, the N-terminal Cre fragment was fused to a Coh2 (CreN59-Coh2) domain, and the fusion protein is driven by a constitutive promoter P_{hCMV}. The C-terminal Cre fragment was fused to a DocS domain and this fusion (DocS-CreC60) can be induced by the FRL-responsive promoter P_{FRLx}. Under FRL illumination, the active bacterial photoreceptor BphS can convert intracellular guanylate triphosphate (GTP) into cyclic diguanylate monophosphate (c-di-GMP) which enables the translocation of the hybrid transactivator FRTA (p65-VP64-BldD) into the nucleus and binds to the FRL-responsive promoter P_{FRL} to drive the expression of the fusion domain DocS-CreC. The interaction of Coh2 and DocS domains leads to reinstatement of Cre enzyme catalytic activity, which allows excision of the stop cassette flanked by *loxP* sites, leading to target gene expression.

10.2.1 Optimization of the Different FRL-Responsive Promoters P_{FRLx}

10.2.1.1 Construction of Three Different FRL-Responsive Chimeric Promoters

To reduce the background activity, we first constructed three different FRL-responsive chimeric promoter variants (P_{FRLx}), including P_{FRLa} [(whiG)$_3$-$P_{hCMVmin}$)], P_{FRLc} [(whiG)$_3$-P_{min})], and P_{FRLd} [(whiG)$_3$-TATA)] and tested the different promoter configurations for driving DocS-CreC60 expression. Then, we optimized different amounts of the plasmids encoding the CreC60 fusion fragments and found that P_{FRLd} [(whiG)$_3$-TATA)] driving DocS-CreC60 exhibited the highest FRL-triggered DNA recombination efficiency that was evaluated using the SEAP reporter assay.

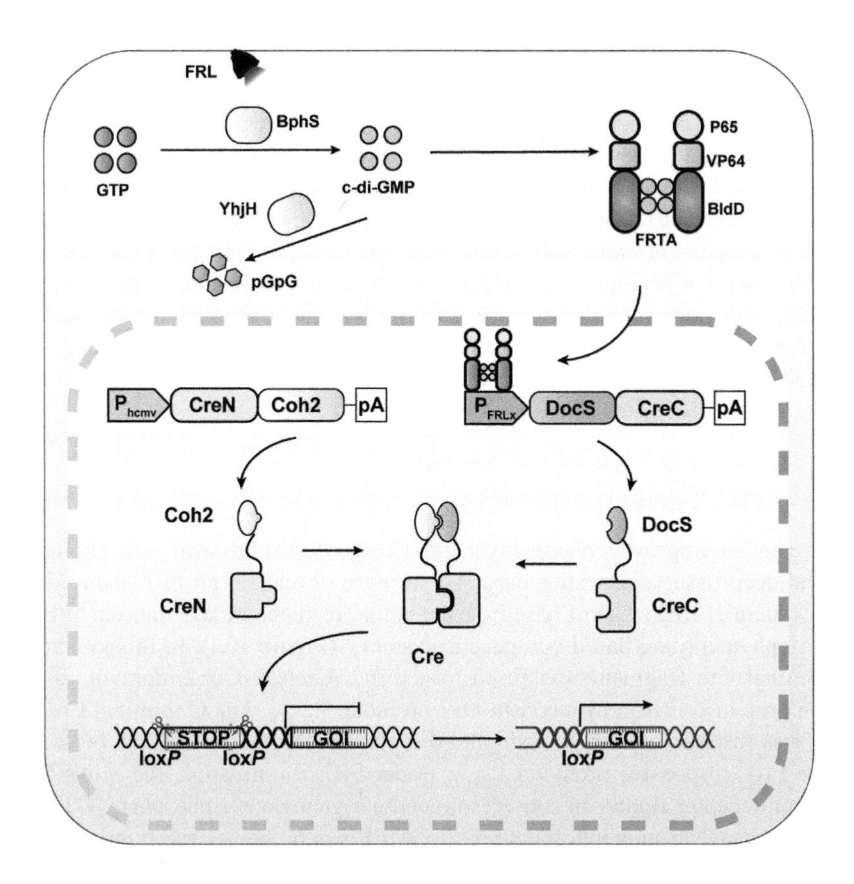

FIGURE 10.1 Schematic representation of the FISC system. Cre recombinase was split into two fragments: CreN59 fused to Coh2 driven by a constitutive promoter (P_{hCMV}) and CreC60 fused to DocS driven by the far-red light (FRL, 730 nm)-inducible promoter (P_{FRLx}). Upon FRL illumination, the photoreceptor BphS is activated to convert intracellular guanylate triphosphate (GTP) into cyclic diguanylate monophosphate (c-di-GMP). The cytosolic c-di-GMP production induces the binding of the far-red light-dependent transactivator FRTA (p65-VP64-BldD) to its synthetic promoter P_{FRLx} to drive DocS-CreC60 expression. Consequently, the catalytic activities of Cre recombinase can be restored once the two Cre fragments assemble based on affinity interactions of their respective Coh2 and DocS fusion domains, enabling to excise DNA sequences flanked by *loxP* sites.

10.2.1.2 Step-by-Step Showcase Protocol

Cell transfection

1. Seed 6×10^4 HEK293 cells per well in a 24-well cell culture plate and culture for 18 h.
2. Transfect each well (24-well plate) with the plasmid DNA for the FISC system, including the FRL-responsive sensor pXY137 (P_{hCMV}-p65-VP64-BldD-pA::P_{hCMV}-BphS-P2A-YhjH-pA, 100ng), pXY169 (P_{hCMV}-CreN59-L9-Coh2-NES-pA, 10 ng), pXY177 (P_{FRLd}-NLS-DocS-L9-CreC60-pA, 10 ng) and SEAP reporter pGY125 (P_{hCMV}-*loxP*-STOP-*loxP*-SEAP-pA, 200 ng) diluted in 50 µL of FBS-free and antibiotic-free DMEM medium.

3. Add 0.96 μL polyethyleneimine (PEI, 1 μg/μL, PEI and DNA at a ratio of 3:1) and mix adequately.
4. Incubate the 50 μL DNA-PEI mixture solution at room temperature for 15 min to allow complex formation between the positively charged PEI (amine groups) and the negatively charged pDNA (phosphate groups) and add dropwise to the cells.
5. Change the medium with fresh DMEM medium supplemented with 10% (vol/vol) fetal bovine serum (FBS) and a 1% (vol/vol) penicillin and streptomycin solution 6 h after transfection.

FRL illumination

1. Place the culture plate below a custom-designed 4 × 6 far-red LED array 24 h after transfection (Figure 10.2A and 2B).
2. Manually adjust the light illumination time and intensity.
3. Expose cells to FRL LED (1.5 mW cm^{-2}, 730 nm) for 6 h each day for two days. The dark samples were covered with an aluminum foil protecting these from ambient light illumination.

SEAP reporter assay

1. Collect cell culture supernatant (200 μL) at 48 h after the initial illumination.
2. Heat-inactivate cell culture supernatant at 65°C for 30 min.
3. Prepare 2 × SEAP buffer [20 mM of homoarginine, 1 mM $MgCl_2$, 21% (v/v) diethanolamine]. Adjust the pH value to 9.8 and store it away from light at 4°C.
4. Prepare substrate solution containing 120 mM of *p*-nitrophenylphosphate and store it away from light at –20°C.
5. Add 100 μL 2 × SEAP buffer and 20 μL substrate solution to 80 μL heat-inactivated cell culture supernatant.
6. Measure light absorbance at 405 nm at 37°C for 30 min using a multi-mode microplate reader (BioTek Instruments, Inc.).
7. Calculate the SEAP production from the slope of the time-dependent increase in light absorbance.

A

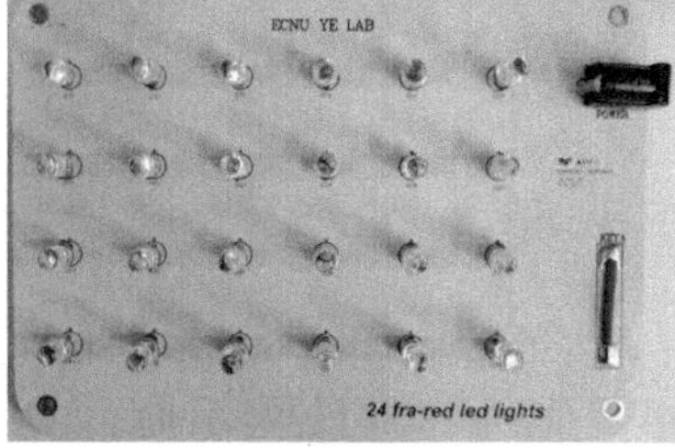

B

FIGURE 10.2 An illumination apparatus used in the study. (A) The 24-channel far-red LED array illumination device. (B) Photograph of the custom-designed far-red LED array used for FISC-controlled transgene expression in mammalian cells grown in a 24-well plate.

10.2.2 Optimization of Different Linkers Between CreN59 and Coh2 as well as Between CreC60 and DocS

To obtain robust DNA recombination efficiency using this system, we further constructed different linkers, including short linkers, semi-flexible linkers, and flexible linkers. We optimized amino acid sequences of these linkers between CreN59 and Coh2 as well as between CreC60 and DocS because protein steric hindrance would impede the reconstitution of Cre activity[12, 19]. After testing the light-inducible recombination efficiency of different combinations of these linkers, we found that one of the 11 (L9/L9) combinations yielded the lowest background recombination level of SEAP in the dark and the highest fold induction upon FRL illumination.

10.2.3 Optimization of the CreN59/CreC60 Catalytic Activity for Different *loxP* Mutants

We constructed different plasmids encoding *loxP* mutants, such as *lox2722*, *lox66*, *lox71*, *lox72*, *JTZ17*, *JTZ17/JT15*, *JT15* and compared CreN59/CreC60 in terms of their catalytic activity for various *loxP* mutants and the original (non-mutant) *loxP*. To test and optimize split Cre fragments, we generated SEAP reporters flanked by different *loxP* mutants (floxed-STOP SEAP). Using the SEAP reporters, we investigated the reassembly of each pair induced by FRL and found that the CreN59/CreC60 pair showed the highest DNA recombination efficiency for the original *loxP*-flanked reporter, whereas no catalytic activity for other *loxP* mutants-flanked reporter sequences (Figure 10.3).

10.3 DNA Recombination Performance of the FISC System in Mammalian Cells

To demonstrate the wide applicability and compatibility of the FISC system, we then transfected the FISC system into different mammalian cell lines and found that it was functional in a number of mammalian cell types. However, DNA recombination efficiency is variable, probably because of different transfection efficiencies, as well as potential interactions with endogenous cell components of different cell lines[22, 24, 25]. Moreover, the FISC system exhibited both illumination-intensity and exposure-time-dependent DNA recombination and precise spatiotemporal Cre activation (Figure 10.4A and 4B).

10.3.1 Step-by-Step Showcase Protocol

10.3.1.1 Exposure-Time and Illumination-Intensity-Dependent FISC System Activity

1. Transfect each well (24-well plate) with the plasmid DNA for the FISC system, including pXY137 (100 ng), pXY169 (10 ng), pXY177 (10 ng), and pGY125 (SEAP reporter plasmid, 200 ng).
2. Place the 24-well culture plate below a custom-designed 4 × 6 far-red LED array at 24 h after transfection.

Step 1: FRL-responsive chimeric promoter

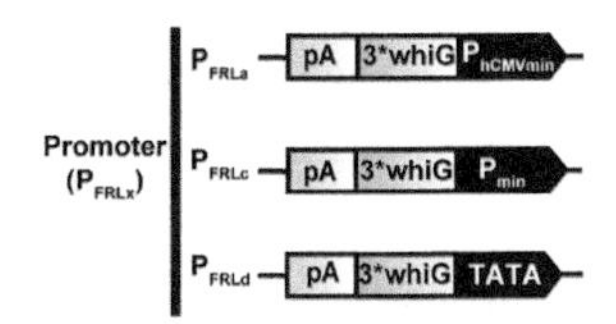

Step 2: Different amounts of the plasmids encoding the CreC60 fusion fragments and CreN59 fusion fragments

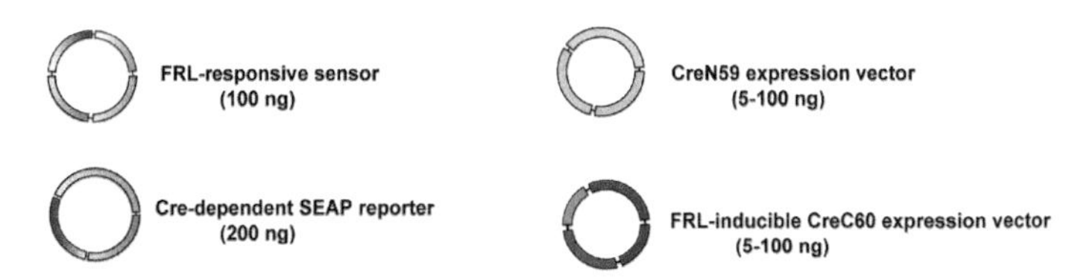

Step 3: Linker amino acid sequences

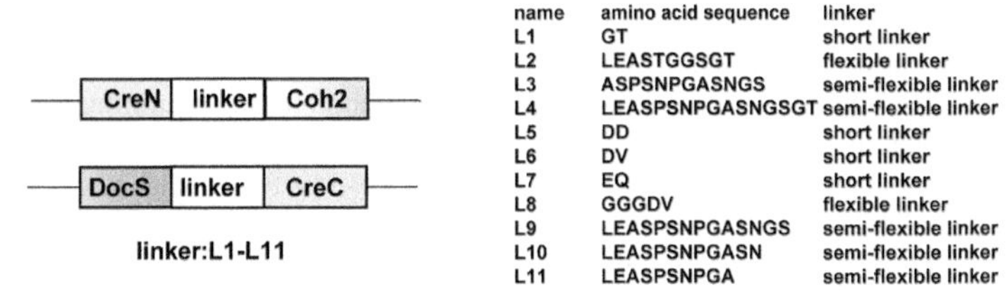

name	amino acid sequence	linker
L1	GT	short linker
L2	LEASTGGSGT	flexible linker
L3	ASPSNPGASNGS	semi-flexible linker
L4	LEASPSNPGASNGSGT	semi-flexible linker
L5	DD	short linker
L6	DV	short linker
L7	EQ	short linker
L8	GGGDV	flexible linker
L9	LEASPSNPGASNGS	semi-flexible linker
L10	LEASPSNPGASN	semi-flexible linker
L11	LEASPSNPGA	semi-flexible linker

Step 4: Different loxP mutants

	Left arm	Spacer	Right arm
loxP:	ATAACTTCGTATAA	TGTATG	CTATACGAAGTTAT
lox2722:	ATAACTTCGTATAA	AGTATG	CTATACGAAGTTAT
lox66:	TACCGTTCGTATAA	TGTATG	CTATACGAAGTTAT
lox71:	ATAACTTCGTATAA	TGTATG	CTATACGAACGGTA
lox72:	TACCGTTCGTATAA	TGTATG	CTATACGAACGGTA
JTZ17:	ATAAATTGCTATAA	TGTATG	CTATACGAAGTTAT
JTZ17/JT15:	ATAAATTGCTATAA	TGTATG	CTATACGAATAATT
JT15:	ATAACTTCGTATAA	TGTATG	CTATACGAATAATT

Step 5: Genetic configuration of the FISC system

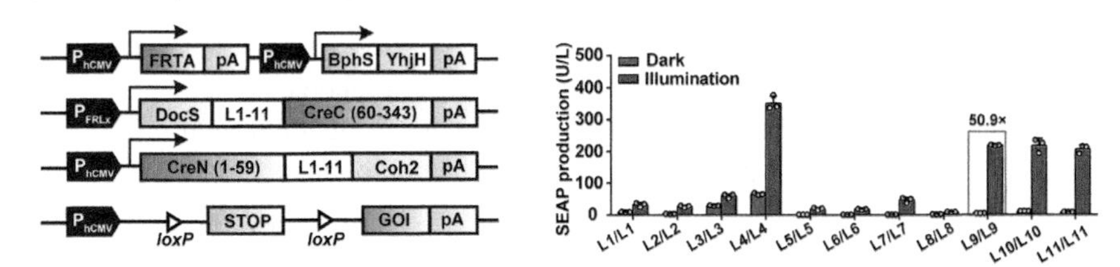

FIGURE 10.3 Schematic representation of multiple strategies for improving the performance of the FISC system. Firstly, different FRL-responsive chimeric promoter (P_{FRLx}) configurations for driving DocS-CreC60 expression were designed and tested in HEK293 cells (Step 1). Secondly, different amounts of the plasmids encoding the CreC60 fusion fragments and CreN59 fusion fragments were optimized (Step 2). Thirdly, different linkers (L1-L11) between Coh2 and CreN59 as well as between DocS and CreC60 were optimized and tested in this study (Step 3). Fourthly, the catalytic capacity of the CreN59/CreC60 pair in various *loxP* mutants was tested (Step 4). Finally, the FISC system with the best performance was developed through the step-by-step improvements. Schematic depicting the genetic configuration of constructs used in the FISC system (Step 5, left), and the FISC system showed the highest fold induction (50.9-fold induction) upon FRL (Step 5, right).

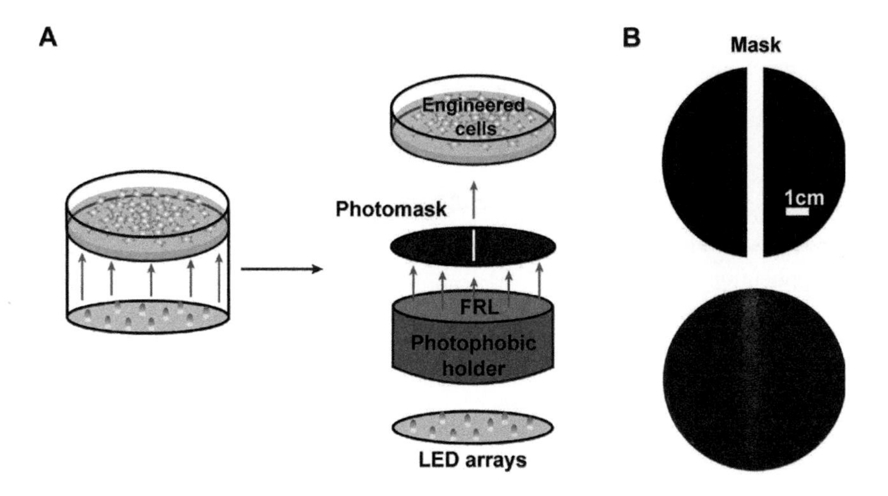

FIGURE 10.4 Evaluation of the spatial resolution for FISC-mediated transgene expression. (A) Setup for pattern illumination. The far-red LEDs were mounted on the bottom of a photophobic holder, and cells were illuminated through a photomask containing a 6.5-mm line pattern. (B) Spatial control of FRL-dependent transgene expression mediated by the FISC system. A monolayer comprising HEK-293 cells was co-transfected with pXY137, pXY169, pXY177, and pDL78 (EGFP reporter plasmid) at a ratio of a 10:1:1:20 (w/w/w/w), illuminated for 6 h with FRL (1.5 mW cm^{-2}, 730 nm) each day for two days, but through a photomask (schematic, top with a 6.5 mm slit, and fluorescence microscopy-based analysis of the corresponding pattern of EGFP expression at 48 h after the first illumination (bottom)).

3. Adjust the different light time (0–120 h) at FRL intensity (1.5 mW/cm^2) or the different light time intensities (0 to 5 mW/cm^2) at 6 h/day for 2 days.
4. Quantify SEAP reporter expression levels at corresponding time after initial illumination.

10.3.1.2 Setup for Pattern Illumination

1. Mount the far-red LEDs on the bottom of a photophobic holder.
2. Manufacture a 6.5-mm line pattern photomask using aluminum foil.
3. Place the patterned photomask on top of the photophobic holder.

10.3.1.3 Spatial Control of FRL-Dependent DNA Recombination Mediated by the FISC System

1. Seed 3.5×10^6 cells into a 10-cm cell culture dish.
2. Transfect a 10-cm dish with 16 μg of total plasmid for FISC containing pXY137 (5 μg), pXY169 (0.5 μg), pXY177 (0.5 μg), and pDL78 (EGFP reporter plasmid, 10 μg) using PEI-based transfection reagent.
3. Expose the transfected cells to FRL LED (1.5 mW/cm^2, 730 nm) for 6 h each day with a slit-patterned photomask at 24 h after transfection.
4. Take fluorescence images at 48 h after the first illumination using Clinx imaging equipment (ChemiScope 4300Pro).

10.4 OPTOGENETIC DNA RECOMBINATION IN TDTOMATO TRANSGENIC MICE

10.4.1 *In vivo* DNA Recombination with the FISC System Using Hydrodynamic Injection Delivery

After confirming that the FISC system enables efficient DNA recombination upon FRL illumination *in vitro*, we test whether our FISC system could be used for *in vivo* applications. We first simplified the system by concatenating the constructs for CreN-Coh2 and DocS-CreC60 fusion fragments into a single plasmid pXY237 (pA-CreC60-L9-DocS-NLS-P_{FRLd}-Space3-P_{hCMV}-CreN59-L9-Coh2-NES-P2A-ZeoR-pA). Then we delivered the FRL-responsive sensor pXY137 (P_{hCMV}-p65-VP64-BldD-pA::P_{hCMV}-BphS-P2A-YhjH-pA, μg) and pXY237 (100 μg) plasmids by hydrodynamic injection via the tail vein and followed by illumination with FRL (20 mW cm^{-2}, 730 nm) for 12 h (15 min on, 15 min off, alternating) after 8 h injection (Figure 10.5). We found that tdTomato expression was significantly induced in the mice liver under FRL illumination compared to dark control mice. These results indicated our FISC system enables efficient *in vivo* DNA recombination under FRL light illumination.

10.4.2 *In vivo* DNA Recombination with the FISC System Using AAV Delivery

To further verify whether our FISC system could be used for *in vivo* DNA recombination, we firstly deployed our FISC system containing two AAV vectors encoding light-sensing module BphS and CreN59-Coh2 fragment driven by a constitutive promoter P_{hCMV}, respectively, and one AAV vector encoding hybrid transactivator p65-VP64-BldD driven by a constitutive promoter P_{hCMV} and Docs-CreC60 induced by the FRL-responsive promoter (Figure 10.6A and 6B). Then we transduced these three AAVs into tdTomato transgenic mice via tail vein injection. Two weeks after the injection, the mice were exposed to FRL (20 mW cm^{-2}, 730 nm) for 12 h (15 min on, 15 min off, alternating) each day for two days (Figure 10.7). Seven days after the first illumination, the tdTomato signal from the isolated liver was detected using *IVIS* Lumina II *in vivo* imaging system, and fluorescence imaging indicated that mice under FRL illumination exhibited an about 18-fold increase compared to

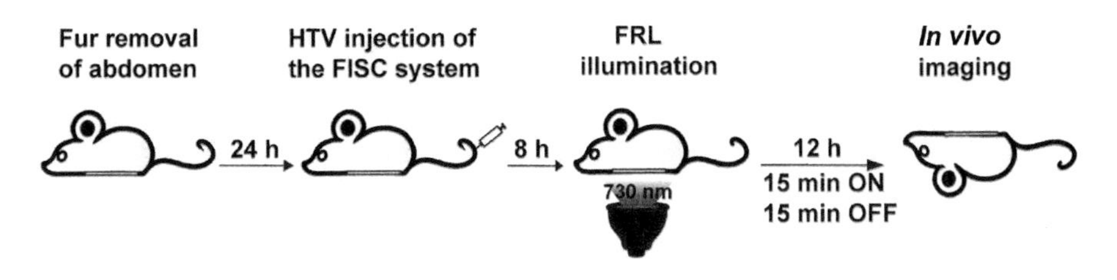

FIGURE 10.5 Schematic representation of the experimental procedures for FISC-medicated DNA recombination in mice.

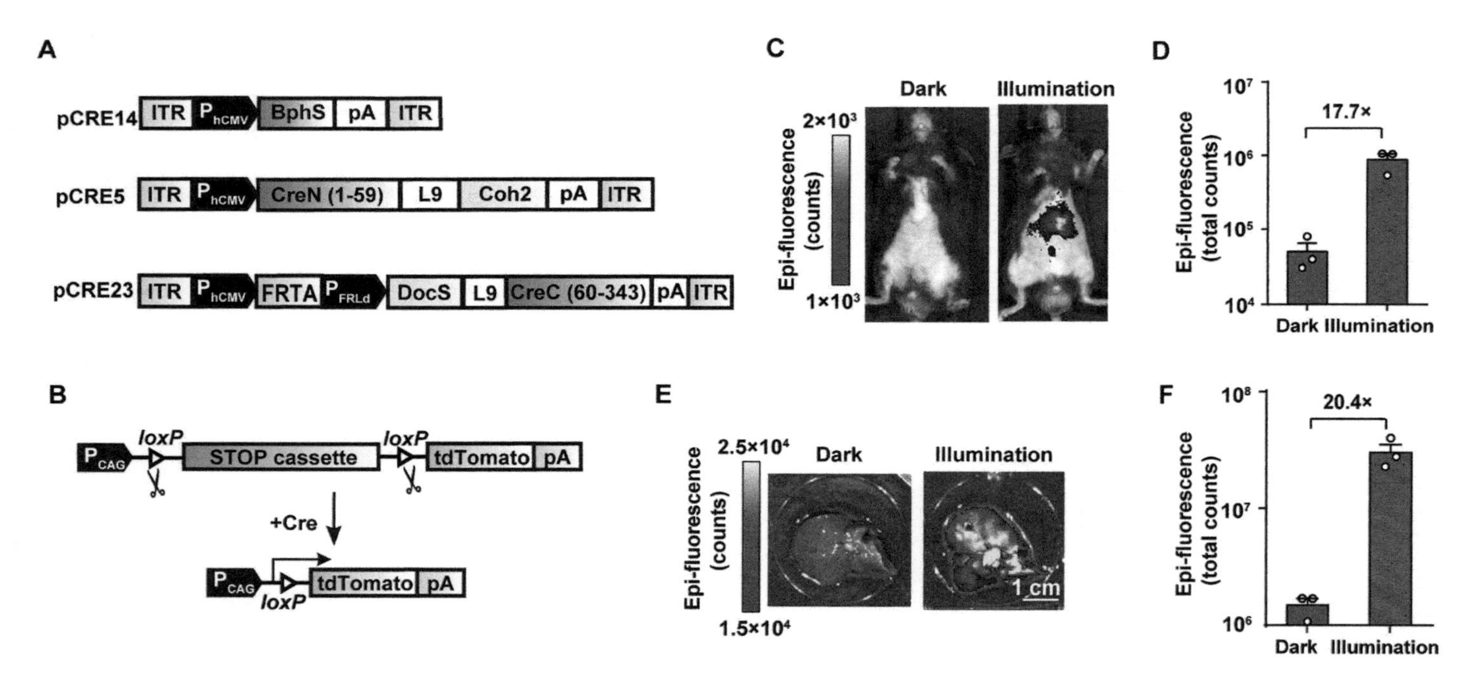

FIGURE 10.6 AAV delivery of FISC-mediated DNA recombination in transgenic Cre-tdTomato reporter mice. (A) Schematic depicting the genetic configuration AAV vectors for the FISC DNA recombination system. (B) Schematic of working principle for transgenic Cre-tdTomato reporter mice in which the *loxP*-flanked STOP cassette can be excised by Cre recombinase to allow Cre reporter (tdTomato) expression. (C) Representative images of the tdTomato fluorescence in the AAV-transducted reporter mice. (D) The fluorescence measurements of the tdTomato expression shown in C. (E) Representative images of the tdTomato fluorescence in isolated liver tissue from the AAV-transduced mice shown in C. Scale bar, 1 cm. (F) Bioluminescence tdTomato expression shown in E. Data in C, E represent the mean ± SEM, $n = 3$ mice/group.

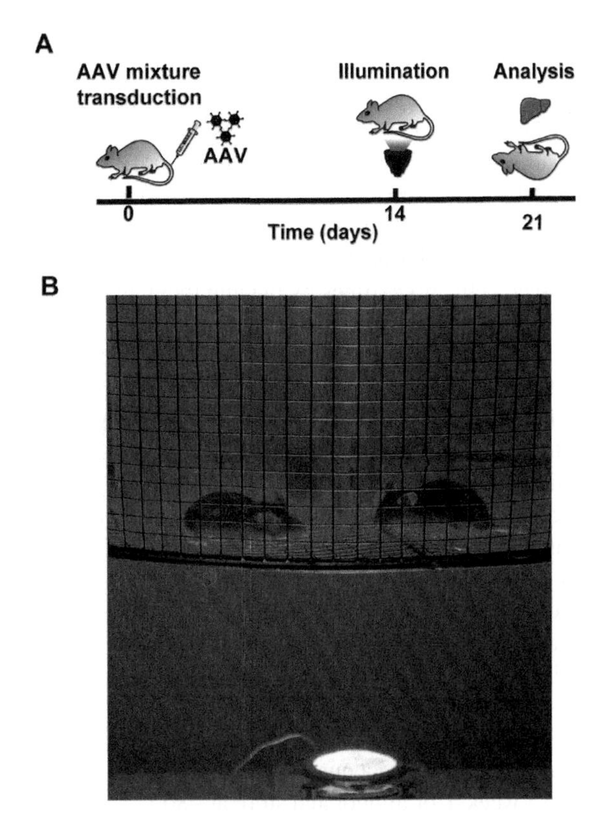

FIGURE 10.7 FISC-mediated DNA recombination in mice. (A) Schematic representation of the experimental procedure for FISC-mediated DNA recombination activity in mice. (B) Mouse illumination device. Mice are illuminated with a LED (50 W, 730 nm, with a light angle of 60°), which was obtained from Shenzhen Kiwi Lighting Co. Ltd. This integrated LED contains 50 small LED (1 W) beads. The light intensity at a wavelength of 730 nm was measured using an optical power meter (Q8230, Advantest).

transduced dark control mice (Figure 10.6C and 6D). Moreover, quantitative analysis of fluorescence imaging revealed an over 20-fold increase in the isolated livers of FRL illuminated mice compared to the transduced dark control mice (Figure 10.6E and 10.6F). These results indicated that the AAV-medicated FISC system showed efficient DNA recombination activity under FRL illumination, which offers a potential approach for constructing Cre transgenic mouse models to achieve spatiotemporal manipulation of genome recombination *in vivo.*

10.4.3 Step-by-Step Showcase Protocol

10.4.3.1 Preparation of AAV-Mediated FISC System

1. Construct packaging plasmid (pRC-2/8), purified helper plasmid (pHelper), and transfer plasmid [pCRE14 (ITR-P_{hCMV}-BphS-pA-ITR), pCRE5 (ITR-P_{hCMV}-

CreN59-L9-Coh2-pA-ITR) and pCRE23 (ITR-P_{hCMV}-FRTA-pA-P_{FRLd}-Docs-L9-CreC60-pA-ITR)].

2. Transfect these plasmids at a 1:1:1 (w/w/w) ratio into HEK-293FT cells plated on a 15-cm dish (8 × 10^6 cells/dish) using PEI.
3. Six hours later, media was replaced with fresh DMEM media containing 10% FBS and 1% penicillin and streptomycin solution.
4. Harvest and resuspend cells in 3 ml lysis buffer (20 mM Tris-HCl pH 8.0, 150 mM NaCl) after 72 h transfection.
5. Freeze and thaw cells three times by placing them alternately in a dry ice/ethanol bath and 37°C water bath.
6. Add 3 μL of 1M $MgCl_2$ (final concetration, 1 mM) and 75 U benzonase (final concetration, 25 U/ml) into the cell lysis mixture.
7. Incubate cell lysis mixture at 37°C for 40 min and centrifuge at 4000 × g for 10 min to collect the supernatant.
8. Prepare iodixanol gradient solution (see Table 10.1).
9. Overlay each solution into a QuickSeal tube in the order: (1) 3.5 mL of 60% iodixanol; (2) 3.5 mL of 40% iodixanol; (3) 4 mL of 25% iodixanol; (4) 4 mL of 17% iodixanol), using a 10 mL syringe.
10. Carefully add the virus supernatant on top of the iodixanol gradient
11. Centrifuge at 350,000 × g for 2 h at 4°C.
12. Isolate the 40% iodixanol fraction, wash with PBS and concentrate to a final volume 100–200 μL using Amicon Ultra 15 centrifugal filter devices-100 K (Millipore, Bedford, MA).
13. Titer the purified AAV using a quantitative PCR.
14. Store at 4°C (2 weeks), or aliquot and store at –80°C.

10.4.3.2 FISC-Mediated DNA Recombination in Mice

1. Randomly divide *Gt(ROSA)26Sor$^{tm14(CAG\text{-}tdTomato)Hze}$* mice (male, 8-week-old) into two groups with three mice in each group.
2. Transduce 1 ml AAV virus mixture containing three AAV vectors: AAV-P_{hCMV}-BphS-pA (2 × 10^{11} vg), AAV-P_{hCMV}-CreN59-L9-Coh2-pA (1 × 10^{11} vg) and AAV-P_{hCMV}-FRTA-pA-P_{FRLd}-Docs-L9-CreC60-pA (2 × 10^{11} vg) into these mice via tail vein injection.

TABLE 10.1
Preparation of Iodixanol Gradient Solution

Iodixanol gradient	10 × PBS (mL)	1 M $MgCl_2$ (mL)	1 M KCl (mL)	5 M NaCl (mL)	Iodixanol (mL)	0.5% Phenol red (mL)	ddH_2O (mL)
17% iodixanol	5	0.05	0.125	10	12.5	–	22.325
25% iodixanol	5	0.05	0.125	–	20	0.1	24.725
40% iodixanol	5	0.05	0.125	–	33.3	–	11.525
60% iodixanol	–	0.05	0.125	–	50	0.025	–

3. Illuminate the mice with LED far-red light (730 nm) at an intensity of 20 mW cm^{-2} for 12 h (15 min on, 15 min off, alternating) each day for two days after two weeks of injection.
4. Anaesthetize the mice with isoflurane 7 days after the first illumination. Detect the tdTomato signal from the isolated liver of the mice using IVIS Lumina II *in vivo* imaging system (Perkin Elmer, USA). Sacrifice the mice and isolate the livers for qPCR, Western Blot, and histological analysis.

Abbreviations: **BldD**, *Streptomyces coelicolor* transcription factor regulating hyphae formation; **BphS**, engineered bacterial diguanylate cyclase; **Coh2**, anchoring proteins of *C. thermocellum*; **Cre**, cyclization recombination enzyme; **EGFP**, enhanced green fluorescent protein; **FRL**, far-red light; **FRTA**, mammalian far-red light-dependent transactivator (p65-VP64-BldD); **FISC**, far-red light-induced split Cre-*loxP* system; **GOI**, gene of interest; **ITR,** inverted terminal repeat; ***loxP***, a 34 bp sequence from P1 phage that Cre recombinase binds (5'-ATAACTTCGTATAGCATACATTATACGAAGTTAT-3'); **P2A**, picornavirus-derived self-cleaving peptide engineered for bicistronic gene expression in mammalian cells; **p65**, 65 kDa transactivator subunit of NF-kB; **pA**, polyadenylation signal; **PFRLx**, BldD-based synthetic mammalian far-red light-inducible promoter variants; $\mathbf{P_{hCMV}}$, human cytomegalovirus immediate early promoter; $\mathbf{P_{hCMVmin}}$, minimal version of P_{hCMV}; $\mathbf{P_{min}}$, minimal eukaryotic promoter; **SEAP**, human placental secreted alkaline phosphatase; **STOP**, a long fragment contains polyadenylation signal to prevent transcription; **TATA,** minimal eukaryotic promoter with only TATA box; **VP64**, tetrameric core of Herpes simplex virus-derived transactivation domain; **whiG**, BldD-specific binding sequence; **YhjH**, bacterial c-di-GMP phosphodiester.

10.4 PERSPECTIVES

The Cre-*loxP* recombination systems are powerful genetic engineering tools to precisely manipulate genomic DNA. An optimal Cre-*loxP* recombination system should have the capacity for precise spatiotemporal control, low leakiness, reduced toxicity and invasiveness, and should be easy to use. Our FISC system has been developed to successfully induce efficient DNA recombination *in vivo* upon illumination with a non-invasive FRL LED light. However, each Cre-*loxP* recombination system has its own pros and cons. In our system, the FISC components are relatively complicated and collectively exceed AAV maximum packaging capacity limit. An ideal solution is to develop an optogenetic system with a simple design and rapid activation/deactivation features in response to light. Note that our group recently reported that the red/far-red light-mediated and minimized ΔPhyA-based photoswitch (REDMAP) system works well for *in vivo* DNA recombination due to its small and highly sensitive features[26]. Advancement of optogenetics toolkits will allow for the creation of an inducible Cre-*loxP* recombination system for highly efficient genetic manipulation *in vivo* with high spatiotemporal precision.

ACKNOWLEDGMENTS

This work was financially supported by grants from the National Key R&D Program of China (no. 2019YFA0904500, no. 2019YFA0110802), the National Natural Science Foundation of China (NSFC: no. 32171414, no. 31971346, no. 31861143016, no.31870861), the Science and Technology Commission of the Shanghai Municipality (No. 18JC1411000), and the Fundamental Research Funds for the Central Universities.

REFERENCES

1. Zomer, A., et al., In Vivo imaging reveals extracellular vesicle-mediated phenocopying of metastatic behavior. *Cell*, 2015. **161**(5): pp. 1046–1057.
2. Roh-Johnson, M., et al., Macrophage-dependent cytoplasmic transfer during melanoma invasion in Vivo. *Dev Cell*, 2017. **43**(5): pp. 549–562 e6.
3. Stauffer, W.R., et al., Dopamine neuron-specific optogenetic stimulation in rhesus macaques. *Cell*, 2016. **166**(6): pp. 1564–1571 e6.
4. Polstein, L.R., et al., An engineered optogenetic switch for spatiotemporal control of gene expression, cell differentiation, and tissue morphogenesis. *ACS Synth Biol*, 2017. **6**(11): pp. 2003–2013.
5. Guo, F., D.N. Gopaul, and G.D. van Duyne, Structure of Cre recombinase complexed with DNA in a site-specific recombination synapse. *Nature*, 1997. **389**(6646): pp. 40–46.
6. Branda, C.S. and S.M. Dymecki, Talking about a revolution: The impact of site-specific recombinases on genetic analyses in mice. *Dev Cell*, 2004. **6**(1): pp. 7–28.
7. Loonstra, A., et al., Growth inhibition and DNA damage induced by Cre recombinase in mammalian cells. *Proc Natl Acad Sci U S A*, 2001. **98**(16): pp. 9209–9214.
8. Utomo, A.R., A.Y. Nikitin, and W.H. Lee, Temporal, spatial, and cell type-specific control of Cre-mediated DNA recombination in transgenic mice. *Nat Biotechnol*, 1999. **17**(11): pp. 1091–1096.
9. Yeh, E.S., et al., Tetracycline-regulated mouse models of cancer. *Cold Spring Harb Protoc*, 2014. **2014**(10): pp. pdb top069823.
10. Metzger, D., et al., Conditional site-specific recombination in mammalian cells using a ligand-dependent chimeric Cre recombinase. *Proc Natl Acad Sci U S A*, 1995. **92**(15): pp. 6991–6995.
11. Schwenk, F., et al., Temporally and spatially regulated somatic mutagenesis in mice. *Nucleic Acids Res*, 1998. **26**(6): pp. 1427–1432.
12. Jullien, N., et al., Regulation of Cre recombinase by ligand-induced complementation of inactive fragments. *Nucleic Acids Res*, 2003. **31**(21): p. e131.
13. Manolagas, S.C., C.A. O'Brien, and M. Almeida, The role of estrogen and androgen receptors in bone health and disease. *Nat Rev Endocrinol*, 2013. **9**(12): pp. 699–712.
14. Link, K.H., Y. Shi, and J.T. Koh, Light activated recombination. *J Am Chem Soc*, 2005. **127**(38): pp. 13088–13089.
15. Inlay, M.A., et al., Synthesis of a photocaged tamoxifen for light-dependent activation of Cre-ER recombinase-driven gene modification. *Chem Commun (Camb)*, 2013. **49**(43): pp. 4971–4973.
16. Edwards, W.F., D.D. Young, and A. Deiters, Light-activated Cre recombinase as a tool for the spatial and temporal control of gene function in mammalian cells. *ACS Chem Biol*, 2009. **4**(6): pp. 441–445.
17. Kennedy, M.J., et al., Rapid blue-light-mediated induction of protein interactions in living cells. *Nat Methods*, 2010. **7**(12): pp. 973–975.

18. Taslimi, A., et al., Optimized second-generation CRY2-CIB dimerizers and photoactivatable Cre recombinase. *Nat Chem Biol*, 2016. **12**(6): pp. 425–30.
19. Kawano, F., et al., A photoactivatable Cre-loxP recombination system for optogenetic genome engineering. *Nat Chem Biol*, 2016. **12**(12): pp. 1059–1064.
20. Müller, K., et al., A red/far-red light-responsive bi-stable toggle switch to control gene expression in mammalian cells. *Nucleic Acids Res*, 2013. **41**(7): p. e77.
21. Gomelsky, M., Photoactivated cells link diagnosis and therapy. *Sci Transl Med*, 2017. **9**(387): eaan3936.
22. Shao, J., et al., Smartphone-controlled optogenetically engineered cells enable semiautomatic glucose homeostasis in diabetic mice. *Sci Transl Med*, 2017. **9**(387).
23. Barak, Y., et al., Matching fusion protein systems for affinity analysis of two interacting families of proteins: the cohesin-dockerin interaction. *J Mol Recognit*, 2005. **18**(6): pp. 491–501.
24. Bai, P., et al., A fully human transgene switch to regulate therapeutic protein production by cooling sensation. *Nat Med*, 2019. **25**(8): pp. 1266–1273.
25. Yin, J., et al., A green tea-triggered genetic control system for treating diabetes in mice and monkeys. *Sci Transl Med*, 2019. **11**(515): pp. eaav8826.
26. Zhou, Y., et al., A small and highly sensitive red/far-red optogenetic switch for applications in mammals. *Nat Biotechnol*, 2022. **40**(2): pp. 262–272.

11 Tools and Technologies for Wireless and Non-Invasive Optogenetics

Guangfu Wu, Vagif Abdulla, Yiyuan Yang, Michael J. Schneider, and Yi Zhang

CONTENTS

11.1 Introduction 151
11.2 Wireless Passive Optogenetic Devices and Power Delivery Mechanisms 152
11.3 Wireless Active Optogenetic Devices and Communication Technologies 158
11.4 Wireless Multimodal Optogenetics 160
11.5 Upconversion Nanomaterials and *in vivo* Non-Invasive Optogenetics 163
11.6 Mechanoluminescence and Sono-Optogenetics 167
11.7 Conclusion 170
Acknowledgment 172
References 172

11.1 INTRODUCTION

Optogenetics is an emerging biotechnology that integrates genetic engineering and optical techniques to precisely manipulate neurons of interest[1–3]. This technology can specifically fire or silence the targeted neurons with high spatiotemporal resolution, making it a powerful tool for analyzing neural circuits and their relationships with outcome behavior. The traditional technology relies on rigid optical fibers adapted from the telecommunication industry[4], which could cause damage to the brain tissue and physically constrain freely moving animals. Hence, wireless non-invasive optogenetic technologies that can mitigate these physical damages and constraints while delivering light to target tissue regions and cell populations are attracting more attention. This chapter starts with the discussion of wireless passive optogenetic devices and power delivery mechanisms. It also discusses wireless active optogenetic devices and communication technologies followed by wireless multimodal optogenetics. It continues with the discussion of upconversion nanomaterials and *in vivo* non-invasive optogenetics. Finally, this chapter includes recent developments in mechanoluminescence and sono-optogenetics.

DOI: 10.1201/b22823-11

11.2 WIRELESS PASSIVE OPTOGENETIC DEVICES AND POWER DELIVERY MECHANISMS

Advancements in microfabrication techniques in the semiconductor industry have led to the ultra-miniaturization of many commercially available electronic components, integrated circuits (ICs), and other systems. Among these advancements, microscale light-emitting diodes (μLEDs)[5] are of particular interest because they present unique opportunities for light delivery mechanisms within the context of optogenetics. These μLEDs can be manufactured to be comparable in size to the soma of individual neurons, allowing precise localized optical modulation over individual cells[6]. This technique eliminates all forms of tethered connections previously required for optical waveguides, laying the foundation for the development of wireless optogenetic systems.

Early wireless optogenetic systems can control the ON/OFF operation of implanted μLEDs but lack independent control of major optogenetic stimulation parameters (e.g., frequency, duty cycle, intensity, wavelength, etc.), real-time programmability, and per-device addressability within a cluster. We refer to these devices as wireless passive optogenetic devices, which can be further categorized into two major classes based on their power delivery mechanisms. The first class uses batteries to store electrochemical energy[7,8]. This strategy enables convenient power supply for wireless optogenetic stimulators over decent durations. However, batteries based on conventional electrochemical architectures such as lithium-ion add heavy weight and occupy considerable space on the optogenetic devices, as their energy storage capacities are proportional to their volume. For instance, the low-capacity lithium-ion battery used in a study by Kathe et al.[8] weighs 0.2 g (roughly one-fifth of the weight of the entire optogenetic platform) and has a volume of 270 mm^3. This 12 mAh battery only lasts 3.5 hours before replacement or recharge. In addition, batteries generally consist of hazardous materials which raise safety concerns for subdermal implantations. Therefore, batteries are typically placed on an external part of the device that protrudes beyond the skin. Meanwhile, battery-powered optogenetic systems are not suitable for experiments involving freely behaving and socially interacting small animal models, such as mice. They could significantly limit animals' movement and induce anxiety-related behaviors[9–11]. Repeated human interventions for replacing depleted batteries will also lead to anxiety-related behaviors. Furthermore, the bulky nature of battery-based devices prevents them from being implanted in body areas of interest where extensive mechanical movement is present, such as the spinal cord and the peripheral nerves.

The second category of wireless passive optogenetic devices exploits wireless power transfer techniques such as radiative far-field, non-radiative near-field (resonant inductive coupling), and ultrasound power transfer, each with its own benefits in terms of penetration, range, and hardware complexity. Far-field power transmission uses energy-carrying radiative radio waves generated by accelerating charged particles, namely, alternating electric currents in a conductor (i.e., an antenna). Radio frequencies typically range from 3 Hz to 300 GHz in most applications (420 MHz–2.4 GHz in wireless devices for neuroscience[12]), depending on the experimental requirements. Low-frequency radio waves have long wavelengths and hence increased penetration and range, whereas high-frequency radio waves easily reflect off obstacles and thus cannot travel far. The relatively long transmission distances, wide frequency range, and easy generation make far-field power transfer appealing for wireless biomedical devices. One such system is reported by Kim et al.[6], which uses a radio frequency of 910 MHz. Although the head-mounted device can be operated reliably from ~1 m away, the high-gain rigid panel antenna used in this study is quite large (~1 cm in length), requiring a protrusion well beyond the mouse's head. Further research efforts combined with advances in fabrication technologies have improved antenna designs, yielding radio-based systems capable of full subdermal implantation and achieving greatly enhanced energy harvesting capability. The first representative device is reported by Montgomery et al.[13] (Figure 11.1). The stimulator consists of a receiver coil, a

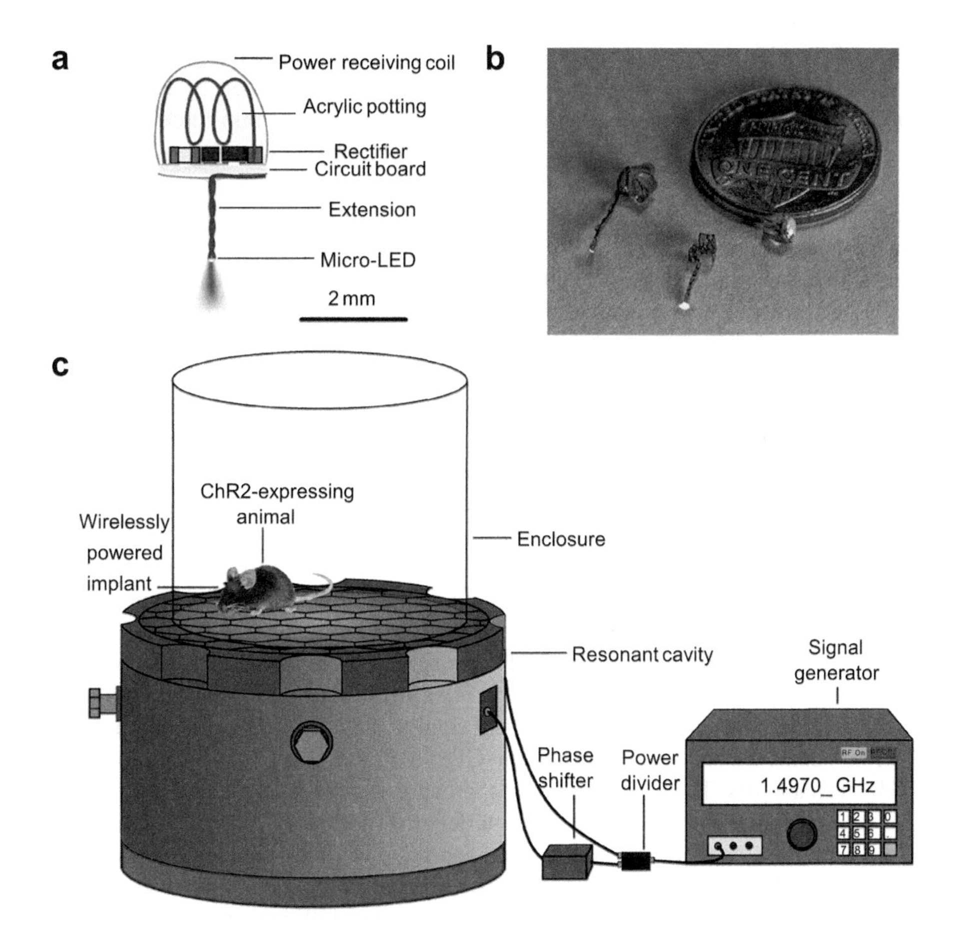

FIGURE 11.1 Subdermal implants for wireless optogenetics. (a) Diagram of the subdermal implant with an extended probe for the brain. (b) Photograph of the peripheral nerve (left), brain (middle), spinal cord (right) implants and their size comparison with a USA one-cent coin. (c) Overview of the wireless optogenetics system. (Reproduced with permission from Ref.[13]; Copyright (2015) Springer Nature.)

two-diode rectification circuit for AC/DC conversion, and a μLED that is connected to the main circuit board through a pair of magnetic wires with lengths determined by the stimulation area of interest (Figures 11.1a and 11.1b). The entire implant is encapsulated in acrylic and, optionally, coated with parylene for enhanced biocompatibility in chronic applications. After implantation, the mice behave freely inside a resonant enclosure where the floor acts as the electromagnetic (EM) transmitter powering the implant at 1.5 GHz (Figure 11.1c).

Despite the small form factor (Figure 11.1b) and reliable optogenetic control, the use of rigid materials in the construction of the reported implantable devices will lead to foreign body responses and scarring during chronic usage due to the constant micromotions at the interface between the probe (Young's modulus ~2.3 GPa[14]) and the neural tissue (Young's modulus 0.1–16 kPa[15–18]), where a huge mechanical mismatch exists. The emergence of soft, flexible, and stretchable materials capable of integration with electronic components provides a promising solution to this problem. One representative wireless optogenetics platform described by Park et al.[19] employs a floating circuit design that is encapsulated using 3 mm-thick polyimide for isolation and, importantly, a 100 μm-thick silicone-based elastomeric layer with low Young's modulus (~0.5 MPa) that makes the entire system mechanically robust and compliant with an effective modulus of ~1.7 MPa (Figure 11.2). All electrical interconnects, including the radio antenna, are defined in stretchable titanium-gold serpentine structures through photolithography (Figure 11.2a). These serpentine traces maintain electrical contact between electronic components (i.e., the rectifier IC and the radio impedance matching circuitry) under repetitive strain deformations, making the device suitable for peripheral nerve and spinal cord stimulation (a soft, stretchable extension to the main device is included in this case for μLED-interfacing within the epidural space) (Figure 11.2b and 11.2c). Furthermore, the use of densely packed serpentine structures and operation at a relatively high radio frequency of 2.34 GHz allows the radio antenna to be space-optimized and fit within an area of 3 × 3 mm^2, while still being able to power the implant at a distance of up to ~0.2 m (Figures 11.2d and 11.2e). The result is a miniaturized soft, flexible, stretchable, and fully implantable optogenetic device capable of chronic behavior studies of the peripheral nerves and the spinal cord in live rodents (Figures 11.2f and 11.2g).

Devices that employ radiative far-field transmission at high frequencies, such as those detailed previously, have a few drawbacks. First, these frequencies are quite common in the telecommunications industry, widely employed by Long-Term Evolution (LTE; 0.41–5.9 GHz), Global Positioning System (GPS; 1.18–1.58 GHz), and WiFi (2.4 and 5.0 GHz) technologies. The abundance of background signals with similar frequencies in the environment, therefore, causes EM interference with the abovementioned optogenetic platforms. Thus, additional hardware and setup for shielding (i.e., Faraday cage) are required for optimal performance. Furthermore, the high-frequency radio waves used in far-field transmission can interfere with their own reflected or scattered components due to interfacial effects, particularly with metal obstructions[20]. Additionally, they cannot penetrate deep into soft biological tissue due to the radiative nature of far-field and strong absorption in tissue[21], resulting in heat accumulation with subsequent perturbations to biological function[22]. This makes radio-based deep wireless implants less viable in larger animal models such as rats, pigs, and non-human primates[12]. Finally,

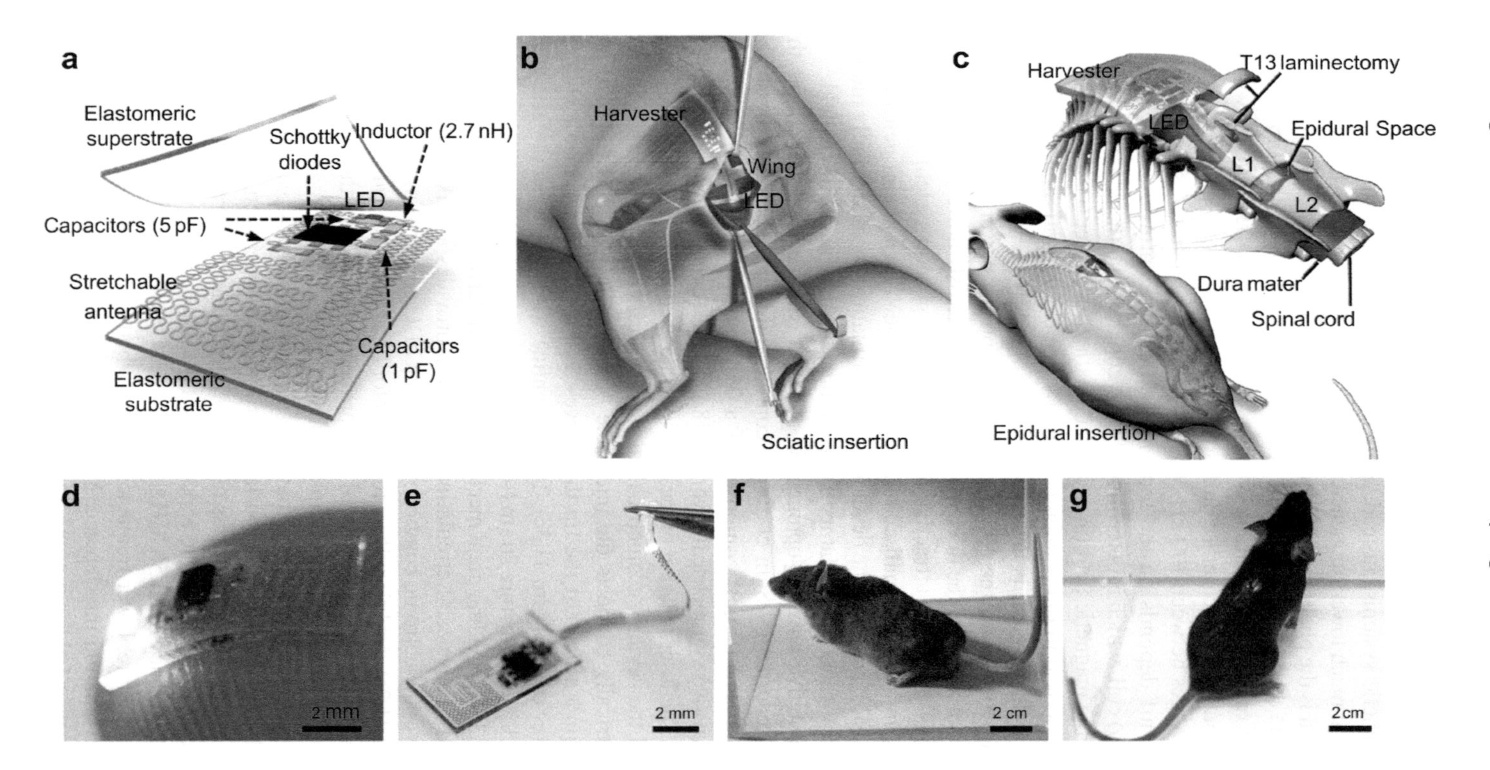

FIGURE 11.2 Soft, flexible, and stretchable wireless optogenetic implants. (a) Exploded view of the device schematic. (b,c) Schematic illustrations of the placement of the implants for peripheral nerve (b) and spinal cord stimulation (c). (d) Photograph of the working optogenetic implant placed on the fingertip. (e) Photograph of the spinal cord implant with the stretchable μLED extension. (f,g) Photograph of the mice implanted with the peripheral nerve (f) and spinal cord stimulators (g). (Reproduced with permission from Ref.[19]; Copyright (2015) Springer Nature.)

far-field transmission efficiency decreases with geometrical misalignment, resulting in significantly reduced angular freedom with high gain antennas. Recently developed wireless optogenetic systems, instead, employ non-radiative near-field transmission, which operates at much lower frequencies.

EM induction is the phenomenon where electrical conductors generate electromotive forces (emf) due to changing magnetic fields. It is widely utilized in near-field wireless power transfer and communication applications in the form of two coils magnetically coupled to each other, in resonance with a target resonance frequency that typically ranges between 100 kHz and 200 MHz (resonant inductive coupling). The induced emf depends on the total magnetic flux captured by the receiver (i.e., secondary) coil, which is determined by the coil area and the angle between the coil plane and the magnetic field lines created by the transmitter (i.e., primary) coil. Therefore, as opposed to radio antennas used in far-field transmission, the size of a near-field EM induction coil is not directly determined by the wavelength. Instead, the physical dimensions of the coil depend mostly on the power requirements and the electrical complexity of the device that is being wirelessly powered. This form of power transfer is especially attractive for wireless optogenetic systems because the receiver coils can be made small and flat and still have sufficient emf generation to operate the μLEDs, enabling the development of miniaturized subdermally implantable wireless optogenetic stimulators that overcome the drawbacks of far-field.

The representative wireless optogenetic platform (Figure 11.3) that adopts this mechanism is reported by Shin et al.[23] A commercially available tri-layer flexible printed circuit board (fPCB) composed of a 75 μm-thick polyimide sheet sandwiched between two symmetrical layers of 18 μm-thick copper serves as a fabrication substrate. The laser ablation process defines the secondary receiver coil, circuit footprints, and the device's geometrical layout (Figures 11.3a and 11.3b). A serpentine wire electrically connects the injectable microneedle with a μLED (470 nm, 220×270×50 μm^3) affixed at the tip that serves as a localized stimulator with the rectified receiver coil (Figure 11.3c). The serpentine wire maintains integrity with strain levels below the fracture strain (~5%) even at extensive 300% uniaxial elongation (Figure 11.3d), providing great mechanical freedom during surgical implantation. The planar EM receiver coil circulating the device provides a maximum power of ~15 mW to the optoelectronic components upon AC/DC conversion with a rectification IC (Figure 11.3a). Additionally, a red indicator LED that sits around the coil replicates the behavior of the implanted μLED for the convenience of the experimenter to observe device operation after implantation (Figure 11.3b). The entire platform is encapsulated with a thin layer of parylene and polydimethylsiloxane (PDMS) to create a soft, biocompatible isolation layer for long-term operation. The EM transmitter, unlike the previous platforms, is simply a loop antenna wrapped around the 30×30 cm^2 enclosure in which the experiments take place, powering the implants at a much lower frequency of 13.56 MHz.

As discussed before, lower frequencies achieve better range and reduced susceptibility to reflection and interference. Moreover, the 13.56 MHz frequency is commonly used in near-field communication (NFC) protocols, implying the potential integration of wireless communication capabilities into these optogenetic systems, which we will discuss in Section 11.3. EM induction power delivery has one major

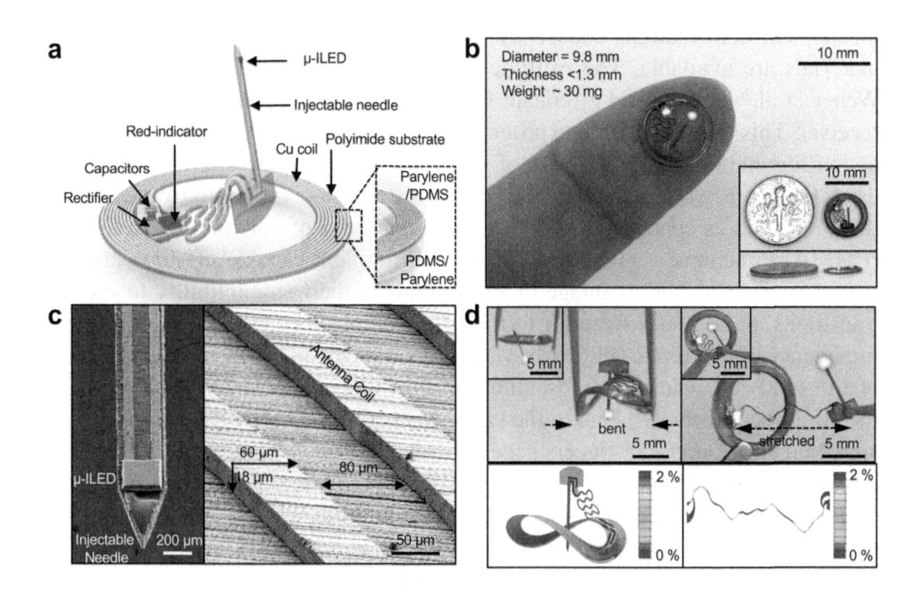

FIGURE 11.3 Flexible planar implants for subdermally implanted wireless optogenetics. (a) Diagram of the wireless optogenetics implant. (b) Photograph of the implant and its size comparisons with a USA one dime coin. (c) Scanning electron microscope (SEM) image of the microneedle tip with the μLED. (d) Finite element modeling results of the bending test for the implant (left) and stretching test for the microneedle serpentine connection (right). (Reproduced with permission from Ref.[23]; Copyright (2017) Elsevier.)

drawback where the power harvesting capabilities of the receiver coil depend heavily on the orientation of its plane with respect to the magnetic field lines (i.e., by a factor of cosθ, where θ is the angle between the field lines and the coil plane). Consequently, when the animals implanted with such devices naturally try to climb the enclosure walls or reach around vertically, the angular mismatch between the transmitter and the receiver causes a significant drop in harvested power, often leading to shut-downs until sufficient alignment is restored. This problem can be mitigated to an extent by implementing multiplexed cage loop antenna designs[24] or incorporating supercapacitors[25] in parallel with the coil antenna to store the harvested energy (charging time of ~30 s) and to sustain the device operation for a certain duration (~1–5 s) in the events of blackouts during large angular mismatches. Additionally, capacitor banks, capable of fast charging (~0.5 s) and discharging (~0.35 ms), can be used to maintain high-intensity optogenetic stimulation modes during reduced power scenarios[10].

Ultrasounds are high-frequency (>20 kHz) sound waves inaudible to humans. For decades, they have been widely used as extremely safe, cost-effective diagnostic tools in medical settings. Due to their low-risk operation and excellent tissue penetration capabilities, ultrasounds are promising for powering next-generation microscale wireless optogenetic deep brain implants that can last a lifetime in models with larger brains such as pigs, non-human primates, and humans. The key component of such systems is a piezoelectric receiver, which generates electrical potentials upon mechanical deformations caused by ultrasound waves. A variety of piezoelectric materials are available. The millimeter-scale optogenetic stimulator described by Weber et al.[26] uses a lead zirconate titanate (PZT) piezoelectric compound as the receiver. This receiver is able to provide ~10 mW of power to the implant for optogenetic illumination with intensities of ~7–15 mW/mm^2, depending on the duty cycle used. On another front, two proof-of-concept studies[27,28] proposed to use zinc oxide (ZnO) nanowires coated in a thin biocompatible layer of poly(methyl methacrylate) (PMMA) to achieve a microscale deep brain implant capable of stimulating individual neurons. A disadvantage of ultrasounds is their susceptibility to interfacial transitions, particularly from air to the tissue. This is because the average acoustic impedance of soft tissue (~1.63 Mrayl) is several orders of magnitude larger than that of air (~400 rayl), causing almost all of the ultrasonic waves to get reflected back into the air at the interface[29]. Due to the strong reflection at the air-tissue interface, conventional ultrasound transducers generally require direct contact with the skin, often with ultrasound gel applied in between. Additionally, the highly directional nature of ultrasound waves requires good geometrical alignment of the transducer with the target device. This can limit the applications of ultrasonically powered deep brain stimulators to only well-controlled, head-fixed scenarios.

11.3 WIRELESS ACTIVE OPTOGENETIC DEVICES AND COMMUNICATION TECHNOLOGIES

All abovementioned passive wireless systems have limited control over optogenetic parameters (i.e., only stimulation frequency and duty cycle) and restricted use in social experiments involving multiple animals due to the lack of individual addressability. In order to expand the capabilities of wireless optogenetic systems, active control components such as microcontrollers (μCs) and system-on-chips (SoCs) for two-way wireless communications such as NFC and Bluetooth can be incorporated. In addition to frequency and duty cycle, these modules can control the devices' stimulation wavelength and intensity, as well as grant them real-time programmability, individual addressability, and sensing capabilities. Consequently, we refer to this general class of devices as wireless active optogenetic devices.

The wide selection of commercially available μCs provides great design flexibility based on the application requirements such as package size, computing capabilities, speed, power consumption, and other features. Integration of a pre-programmed μC to the previously introduced wireless optogenetic platforms (Section 11.2) enables individual control over stimulation parameters for each device inside the same enclosure and multiplexed operation of multiple μLEDs (wavelength control and selective tissue targeting) within the same device. Importantly, all these functionalities are achieved without compromising the device footprint and with minimal bulk added to the system. An array of subdermal systems with said capabilities are reported by Gutruf et al.[24].

Specified by the International Organization for Standardization (ISO), the ISO15693 protocol is a popular NFC standard operating at 13.56 MHz and is widely used to establish communication with contactless integrated circuit cards (i.e., vicinity cards) over relatively large distances[30]. Basic NFC communication using the ISO15693 protocol can be easily achieved by connecting an appropriate NFC SoC directly to the receiver coil of the wireless optogenetic device. The NFC platform is then interfaced with the μC through a serial line, typically using the inter-integrated circuit (I2C) or the serial peripheral interface (SPI) protocols in a master-slave configuration. Both I2C and SPI protocols are common in embedded systems and are used to interface multiple peripheral modules (i.e., slaves) to the same controller (i.e., the master). Upon connection, the desired optogenetic stimulation parameters and operation mode commands can be written wirelessly (over the same transmission antenna used for power delivery) to the memory registers of the NFC chip of the selected device within a group, which are then read by the μC to perform accordingly. In addition to the enhancements discussed previously, the NFC platform further grants the implants real-time programmability and individual device addressability, thereby creating new opportunities for highly complex social behavior experiments. A remarkable study by Yang et al.[11] exploited the capabilities of these implants to facilitate social preference within groups of mice (Figure 11.4). Figures 11.4a and 11.4b display the intensity and wavelength control capabilities of the wireless active implants, in addition to their multiplexed spatial selectivity.

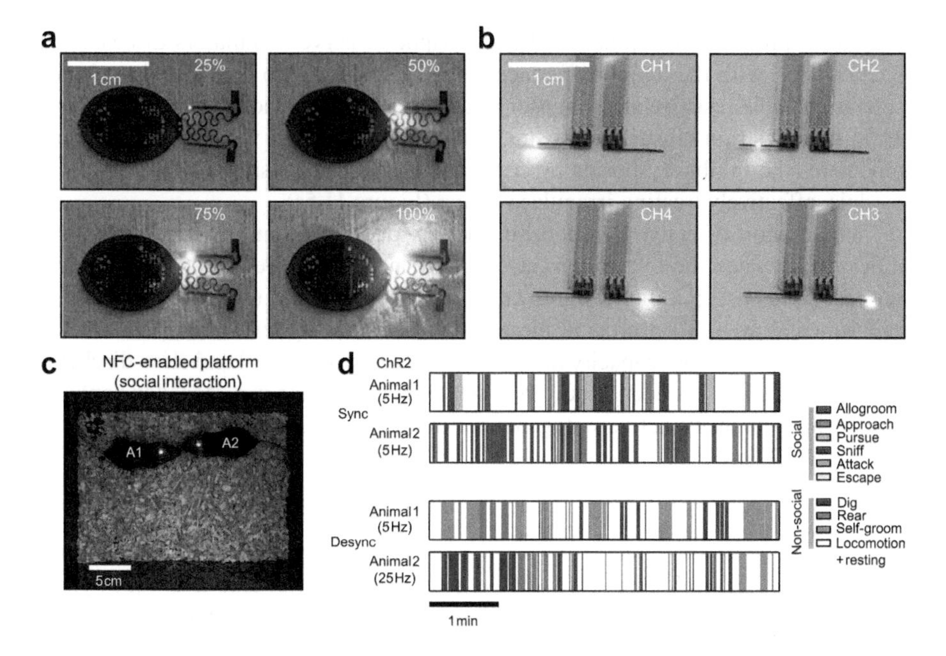

FIGURE 11.4 Wireless active optogenetic devices for studying individual and social behaviors. (a) Intensity control demonstration of the head-mounted wireless optogenetic device. (b) Bilateral wavelength control and spatial selectivity demonstrations of the back-mounted wireless optogenetic device. (c) Social interaction study involving a pair of mice, receiving either synchronized or desynchronized optogenetic stimulations. (d) Social preference (measured in time) results of the experiment based on multiple social behaviors, displaying increased social interaction between the synchronized mice. (Reproduced with permission from Ref.[11]; Copyright (2021) Springer Nature.)

A comprehensive description of the fabrication and the usage of these devices is reported by Yang et al.[31]. Two mice in the experiment received either synchronized (5 Hz tonic for both mice) or desynchronized (5 Hz tonic for one of them and 25 Hz burst for the other) optogenetic stimulation in their medial prefrontal cortex (mPFC) (Figure 11.4c). The result is significantly increased social interaction time between the synchronized pair of mice compared to the desynchronized pair (Figure 11.4d), supporting a recently emerging hypothesis[32,33] that such synchrony in mPFCs may be able to induce social preference.

Bluetooth (BT) is another wireless communication standard that is widely used in consumer electronics. One of its subsets, Bluetooth Low Energy (BLE), is particularly appealing for optogenetic devices. BLE can transmit data efficiently (typically less than half the power consumption of standard BT) at high rates (125 kbit/s, 500 kbit/s, 1 Mbit/s, 2 Mbit/s) over long distances (up to 100 m)[34]. Incorporating BLE SoCs into optogenetic device designs can lead to friendly user interfaces that do not require proprietary hardware and software. Furthermore, the high transmission rate of BLE allows for continuous monitoring using sensors, enabling optogenetic stimulation based on sensed biological signals, which is also referred to as closed-loop optogenetic stimulation. A complete closed-loop system is reported by Mickle et al.[25] for optogenetic peripheral neuromodulation (Figure 11.5). The device is conceptually similar to the spinal cord and peripheral nerve stimulators discussed in Section 11.2, with the major difference being the closed-loop operation based on a soft, stretchable resistive strain sensor wrapped around the mouse bladder (Figures 11.5a-d). The strain sensor detects small and frequent bladder voiding instances and modulates the inhibitory opsins by controlling the μLEDs using a hybrid μC-BLE module, effectively treating irregular voiding (Figure 11.5e).

The addition of active components such as μCs, NFC, and BLE chipsets significantly elevates the overall power requirements of the previous systems, causing increased susceptibility towards angular mismatches and reducing the effective experimental area. To overcome these problems, some systems incorporate wirelessly rechargeable low-capacity batteries that are small in size[35] and slow-discharging supercapacitors as energy buffers[25], while other systems use capacitor banks to realize high-intensity active optogenetic stimulators in an experimental arena as large as 70x70 cm^2 [10].

11.4 WIRELESS MULTIMODAL OPTOGENETICS

Advances in the development of wireless active optogenetics lead to unique hybrid systems with multimodal operation. These systems combine optogenetics with other modalities to extend their capabilities and applications in neuroscience. We have already seen an example of such systems, a hybrid closed-loop device that pairs optogenetics with strain sensing for treating bladder dysfunction in Section 11.3. Another notable example of such multimodal platforms is optofluidics, which combines optogenetics with microfluidics. This class of optofluidic devices incorporates miniaturized electrochemical micropumps that, together with soft microfluidic channels, are capable of delivering drugs[36,37] (Figure 11.6) and neurochemical sampling (microdialysis)[38]. Neuropharmacology modulates neural

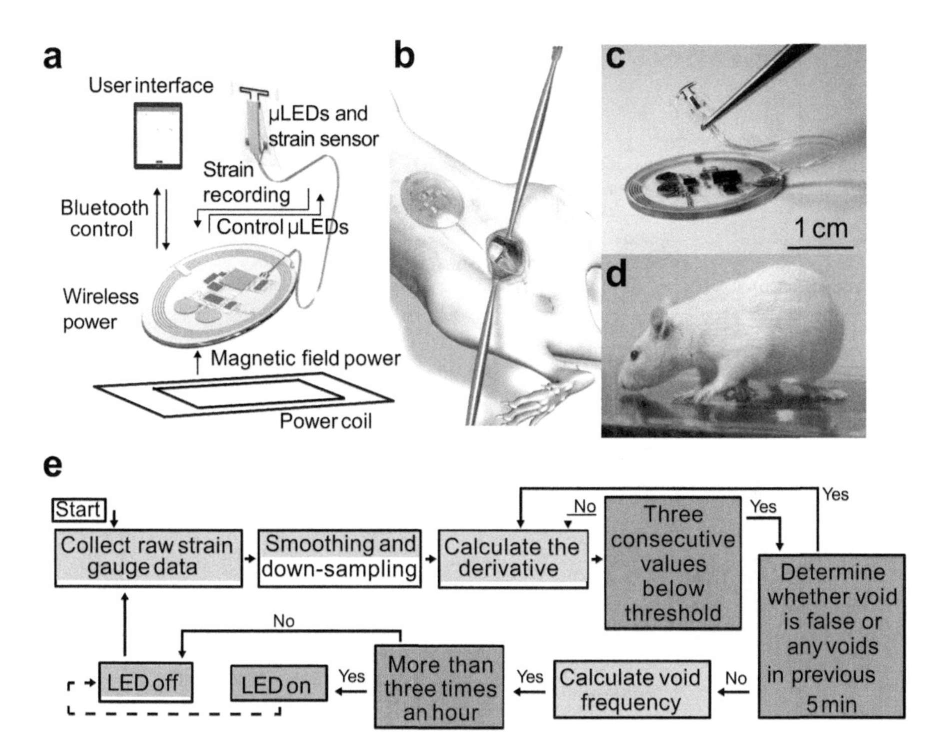

FIGURE 11.5 Active closed-loop wireless optogenetic device for modulating bladder function. (a) System overview of the closed-loop optogenetic device. (b) Schematic illustration of the placement of the device and the strain sensor wrapped around the bladder. (c) Photograph of the soft, implantable closed-loop system. (d) Photograph of the mouse with bladder irregular voiding dysfunction, implanted with the device. (e) The closed-loop optogenetic control scheme for preventing irregular voiding of the bladder. (Reproduced with permission from Ref.[25]; Copyright (2019) Springer Nature.)

activity using drugs that can activate or block cellular function. These drugs are typically infused through a metal cannula, resulting in poor spatiotemporal resolution. Combining neuropharmacology with other modalities, especially optogenetics, is appealing due to the enhanced neural control (Figures 11.6c–e). Potential alternative use for such optofluidic devices is single-step optogenetics, which aims to streamline the surgical procedures involved in optogenetic gene expression and activation by combining viral injection and light delivery into one scheme. Previously, special flexible polymer optical fibers were developed to do this[39]. With the help of emerging optofluidic systems, fully wireless single-step optogenetics can be realized.

The key component that enables ultralow-power microfluidics is the miniaturized electrochemical micropump (Figures 11.6a and 11.6b). The micropump consists of an array of refillable dome-shaped reservoirs that can hold drug or perfusion solution for drug delivery or microdialysis, respectively. The reservoirs are placed on a matching array of cylindrical pump chambers filled with NaOH electrolyte

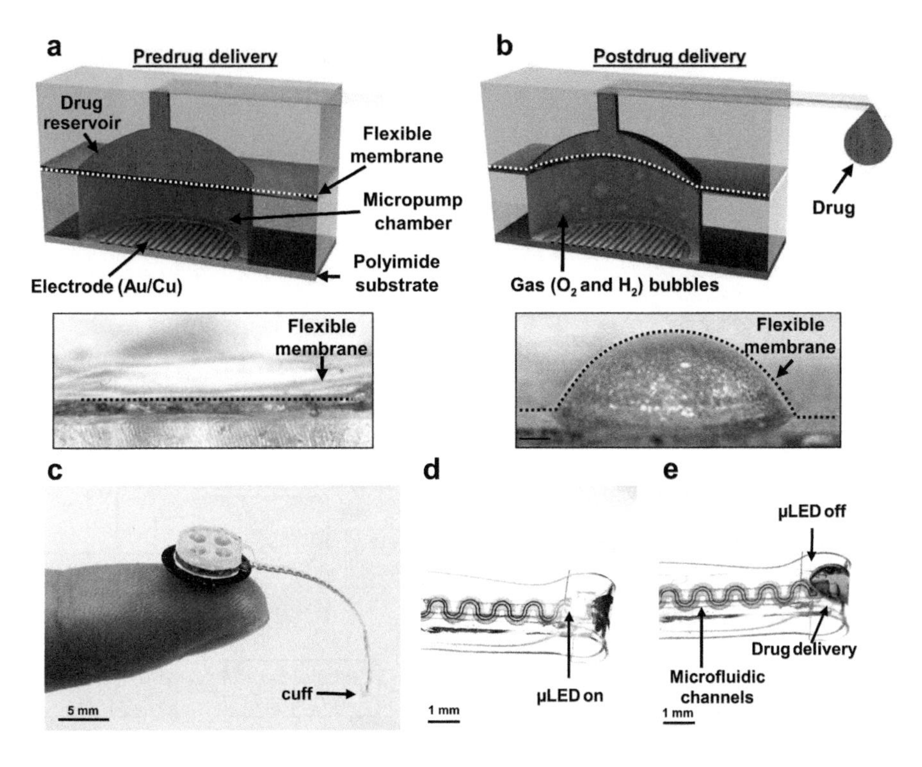

FIGURE 11.6 Wireless optofluidic system for combined optogenetics and neuropharmacology in the peripheral nerves. (a,b) Schematic illustrations of the miniaturized electrochemical micropumps during predrug delivery with resting flexible membrane (a) and postdrug delivery with an expanded flexible membrane (b). (c) Photograph of the wireless optofluidic device with the stretchable extension for optogenetic and neuropharmacological control of peripheral nerves. (d,e) Demonstrations of the cuff µLED (d) and microfluidics channel operations (e). (Reproduced from Ref.[37]; Copyright (2019) American Association for the Advancement of Science (AAAS).)

solution. A flexible membrane of polystyrene-blockpolybutadiene-block-polystyrene (SBS) separates the chamber from the reservoir. Then, the entire structure is placed on the wireless device, where the bottom openings of the micropump chambers align with the corresponding interdigitated electrodes. When voltage is applied across the electrodes using commands from the µC, water electrolysis reactions occur and increase the pressure inside the chambers due to the generation of H_2 and O_2 gas bubbles. The flexible SBS membrane on the top deforms, filling inside the reservoirs and hence pushing the agent through the corresponding microfluidic channels. In the case of neurochemical sampling, the NaOH solution is doped with platinum-nanoparticles (PtNPs) to facilitate recombination reactions of H_2 and O_2 and to reduce the pressure inside, which causes the flexible membrane to recover and pull the neurochemical-bound perfusion solution back into the reservoir.

Other exciting examples of multimodal devices include optogenetic photometry and oximetry. These systems use microscale photodetectors (µPDs) to measure fluorescence signals[9,40] and blood oxygen levels[41] in the surrounding tissue, respectively. In the former, the μLEDs are used to activate genetically encoded calcium indicators (GECIs), whereas the μPDs measure the resulting fluorescence activity. One of the most widely used GECIs is GCaMP which, when bound to calcium ions, emits green light with a peak emission wavelength of 510 nm[42]. As for blood oxygen sensing, the μLEDs illuminate the local tissue, and the μPDs simultaneously measure the differences in intensity levels caused by the unique absorption characteristics of oxygenated and deoxygenated hemoglobin, which is then used to calculate the blood oxygen saturation (SpO_2). These hybrid systems have important implications for future applications, including calcium imaging and implanted tissue oxygen level measurements of internal organs.

11.5 UPCONVERSION NANOMATERIALS AND *IN VIVO* NON-INVASIVE OPTOGENETICS

An optical fiber or μLED probe insertion is required to deliver optical stimulation to the brain tissue because of the limited penetration depth of visible light through the skull and skin. Upconversion nanomaterials have tunable emission wavelengths, narrow bandwidth, and anti-photobleaching properties, making them great candidates for the non-invasive activation of neurons[43–48]. Among many nanomaterials, nanoparticles doped by lanthanide rare earth elements such as Er, Yb, Tm, and Nd are particularly attractive because of their upconversion luminescence properties[43,49,50] (Figure 11.7). Upconversion luminescence refers to the high-energy emission generated by the absorption of two or more photons under the excitation of low-energy photons, also known as anti-Stokes luminescence. Normally, the 4f orbital of the electron shell of lanthanides is not filled with electrons. Thus, the different arrangement of 4f shell electrons makes it possible to generate rich energy levels. The electrons in the 4f shell can absorb and emit light through the transition between energy levels. In addition, the strength of emission can be affected by the dopant level in upconversion materials[44] (Figures 11.7a and 11.7b). For example, lanthanide ions doped host materials have been widely employed to afford efficient UCNPs (upconversion nanoparticles). High lanthanide ions concentration can improve the absorption of excitation light, resulting in strong emission (high-energy emission). The emission spectrum of lanthanides ranges from near-ultraviolet to near-infrared[51] (Figure 11.7c–e). Rare-earth ions have a unique quasi ladder energy level structure and long excited-state lifetime. This allows them to continuously absorb the energy of low-energy excitation, making them more prone to producing upconversion. That is, rare earth ions absorb long-wavelength near-infrared light with low energy and up-convert it into emissions ranging from ultraviolet to near-infrared light[52] (Figure 11.7f), which can mediate the biomedical applications of near-infrared (NIR) light such as upconversion optogenetics[53] (Figure 11.7g).

Due to the low absorption efficiency of NIR light from 700 nm to 1,500 nm in biological tissue, using NIR light as the excitation light source implies relatively deep tissue penetration, creating new opportunities for the applications of NIR in biological tissues. Compared with traditional fluorescent probes such as quantum dots, organic dyes, and fluorescent proteins, upconversion luminescent nanomaterials

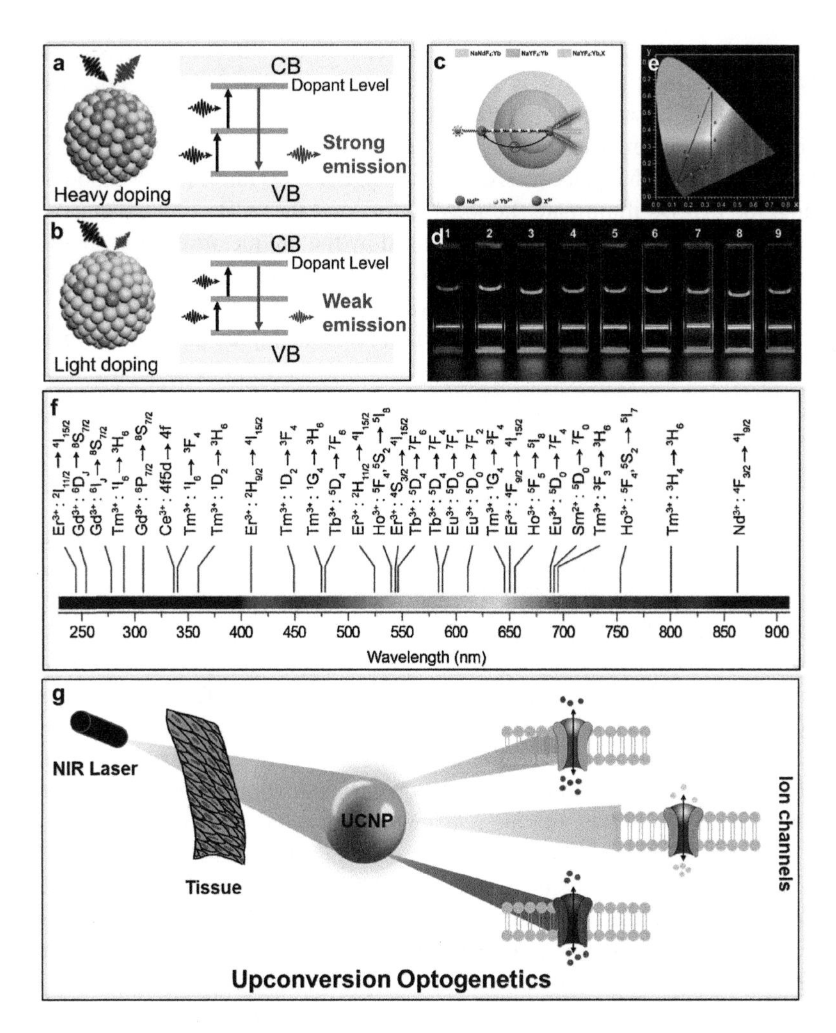

FIGURE 11.7 Working principle of UCNPs and upconversion optogenetics. (a) Illustration of upconversion by heavy doping and (b) light doping. CB: conduction band; VB: valence band. Lanthanide dopant with higher concentration can promote excitation light absorption efficiency. The accumulated low-energy excitation photons can form more high-energy excitation photons (higher CB), which can generate strong upconversion emissions. On the contrary, a lower lanthanide doping level will generate fewer high-energy photons, resulting in weak upconversion emission. (Reproduced with permission from Ref.[44]; Copyright (2020) Elsevier.) (c) Energy-transfer mechanisms in the quenching-shield sandwich-structured UCNPs upon 800 nm excitation. (d) Luminescent photographs of UCNPs with different dopants. (1–3) NaGdF4:Yb,0.5%Tm,xEr@NaGdF4:Yb@NaNdF4:Yb (x = 0, 0.2, and 1%), (4) NaGdF4:Yb,1%Er@NaGdF4:Yb@NaNdF4:Yb, (5–6) NaGdF4:Yb,0.5%Tm,xEr@NaGdF4:Yb,15%Eu@NaNdF4:Yb (x = 1 and 0.2%), (7–9) NaGdF4:Yb,0.5%Tm@NaGdF4:Yb,xEu@NaNdF4:Yb (x = 15, 10 and 5%). (e) CIE diagram of the UCNPs with different dopants as shown in (d). (Reproduced with permission from Ref.[51]; Copyright (2013) John Wiley and Sons.) (f) The electronic transition of lanthanide ions which covers from ultraviolet to near-infrared regions. (Reproduced with permission from Ref.[52]; Copyright (2019) Elsevier.) (g) Design of upconversion optogenetics for ion channels manipulation with NIR light. (Reproduced from Ref[53].)

have good fluorescence quantum yield, photostability, and bleaching resistance[43,54]. More importantly, rare earth upconversion nanomaterials usually use NIR light as the excitation light source, which empowers the imaging of upconversion luminescent nanomaterials with strong penetrability. Furthermore, the optical properties of upconversion nanomaterials greatly promote the application of NIR light in biomedical research such as multimodal imaging of deep tissues, controlled drug release, photothermal therapy, and photodynamic therapy of tumors[55,56]. Therefore, the application of NIR light with relatively deep tissue penetration in optogenetic technology can meet the requirements of delivering light to deep tissues[57].

The UCNP-mediated optogenetics was first proposed by Deisseroth in 2011[58], which became more promising in neuromodulation. This technology was then employed by many research scientists to achieve the activation of light-sensitive proteins via minimally invasive NIR light illumination[59] (Figure 11.8). To date, there are many studies that employ UCNPs-mediated optogenetics to modulate neural activity in neurons[60–62], *Caenorhabditis elegans*[63–65], and zebrafish[66]. In 2015, Hososhima and coworkers designed two upconversion nanocrystals which were applied in upconversion optogenetics[60]. In this study, channelrhodopsin-expressing neurons incubated with upconversion nanocrystals exhibited NIR-mediated photocurrents and action potentials, suggesting the promising potential of upconversion in optogenetics.

In 2016, Bansal et al. demonstrated NIR light-activated optogenetic manipulation *in vitro* and *in vivo* in *Caenorhabditis elegans*[63]. Upconversion blue light triggered by NIR irradiation can fire the mechanical sensory neurons and cause a reversal response akin to being touched, performing Omega turns to avoid light stimulation. In another report, zebrafish was chosen to study the tissue penetration capability of NIR light[66]. UCNPs were incubated with HEK293 cells that can express the channelrhodopsin-2 (ChR2) protein, then injected into zebrafish. Ca^{2+} imaging and caspase-3 detection suggest that 808 nm NIR light can penetrate the zebrafish tissue, remotely activate ChR2 channels and promote Ca^{2+}-mediated apoptosis. Shi group successfully demonstrated UCNP-mediated optogenetics technology in neurons and mice brains[67–69]. The as-prepared UCNPs were coupled with an optical fiber to form a micro-optical probe, which was implanted into the mouse brain. *In vivo* electrophysiology experiments confirm that NIR light (pulse width: 30 ms, power: ChR2, 7 mW/mm^2; C1V1, 4.4 mW/mm^2) irradiation can effectively activate visual cortical neurons[67]. This is a proof-of-concept demonstration of upconversion optogenetics in a rodent model; however, the fiber implant was required to deliver NIR illumination into the mouse brain. Moreover, the previous experiments only support that UCNP-mediated wireless optogenetics technology can regulate neuronal activity *in vivo* at the cellular level. Whether this technology can directly regulate the behavioral changes in rodents is not clear because providing sufficient stimulation to the target tissue of free-moving animals through NIR light wireless irradiation is still a challenge. The same group then designed an automatic projection system, which can automatically identify the position of the mouse head in the open field, track and deliver light in real-time, to ensure that the animals receive effective light stimulation during

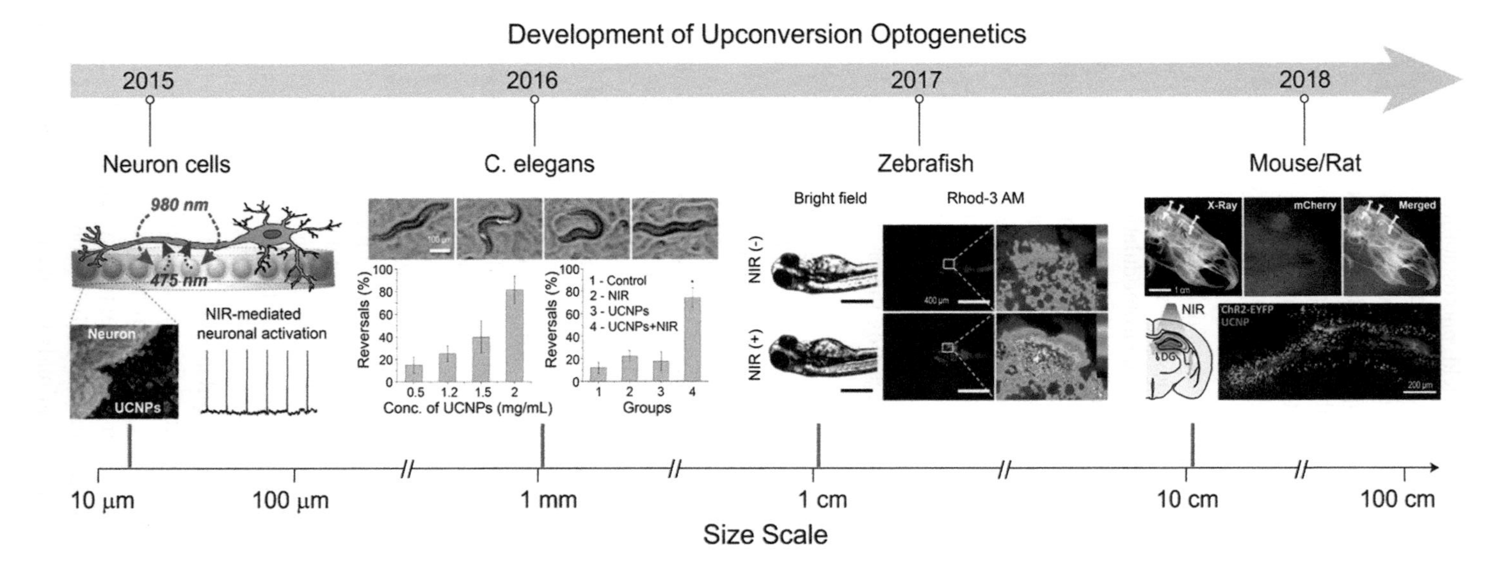

FIGURE 11.8 Representative development of upconversion optogenetics using UCNPs from neuron cells to mouse/rat model. (Reproduced with permission from Ref.[59]; Copyright (2019) American Chemical Society (ACS).)

free movement[68]. In the Y-maze experiment, the animals had a place preference for the arm where the NIR light source was located, suggesting that the NIR light in this area (10 pulses, 20 Hz, 10 ms pulse width, 5 mW/mm^2) effectively activated the dopaminergic neurons in the ventral tegmental area (VTA) to release dopamine transmitters and enhanced the preference of the animals.

In 2018, McHugh et al.[70] used UCNPs-mediated wireless optogenetics technology to induce a variety of physiological and behavioral phenomena in mice, including activating dopaminergic neurons, inhibiting epilepsy, inducing brain oscillation, silencing hippocampal neurons, and triggering memory recall. In this systematic research, NaYF4:Yb/Tm@SiO_2 nanoparticles were injected into the bilateral VTA area (Figure 11.9a-c). After irradiation with a 980 nm NIR laser, it was found that NIR light could activate dopaminergic neurons based on the expression of c-Fos, the electrophysiological activity of neurons, and dopamine level. The expression of c-Fos was imaged by mapping the NIR-mediated dopamine neurons. The percentage of c-Fos-positive neurons in the VTA after NIR stimulation under four different conditions was evaluated by confocal images and summarized in Figure 11.9d. Significantly high populations of c-Fos positive neurons were observed in the brain regions where UCNPs were injected and ChR2 was expressed. NIR-mediated optogenetics was also validated by the observed photocurrent in the neurons in acute slides that were infected with ChR2 in the presence of UCNPs (Figure 11.9e). In addition, transcranial NIR stimulation of VTA dopamine (DA) neurons *in vivo* showed that striatal dopamine was released in the mice with UCNPs injection and ChR2 expression (Figure 11.9f). Furthermore, extracranial low-intensity NIR stimulation significantly reduced the expression of c-Fos protein in excitatory neurons in the hippocampus; that is, the epileptic state of hippocampal neurons was inhibited, suggesting that this technology has great potential in the treatment of neurological diseases.

Overall, UCNP-mediated NIR optogenetics technology can realize minimally invasive regulation of deep brain tissue and overcome the limitation of optical fibers on animal activities and brain tissue damage in traditional technology[53]. Free from the shackles of optical fibers, animals are free to move when stimulated by light and can complete a variety of complex behavioral tests, including open field movement, rotating rod experiment, Y-maze, Skinner box, and other experimental tests.

11.6 MECHANOLUMINESCENCE AND SONO-OPTOGENETICS

Despite the advances of upconversion optogenetics over optical fiber, this method still requires intracranial injection of photoluminescent agents (UCNPs) into the deep brain tissue. It remains challenging to achieve *in vivo* optogenetics in a much less invasive manner. To address the limitations of the current optogenetics technologies, less invasive methods such as intravenous delivery of photoluminescent agents and methods that can trigger photoluminescence with deeper tissue penetration are necessary from the perspective of non-invasive optogenetics[71–75]. Mechanoluminescent (ML) materials produce luminescence under mechanical force, providing new

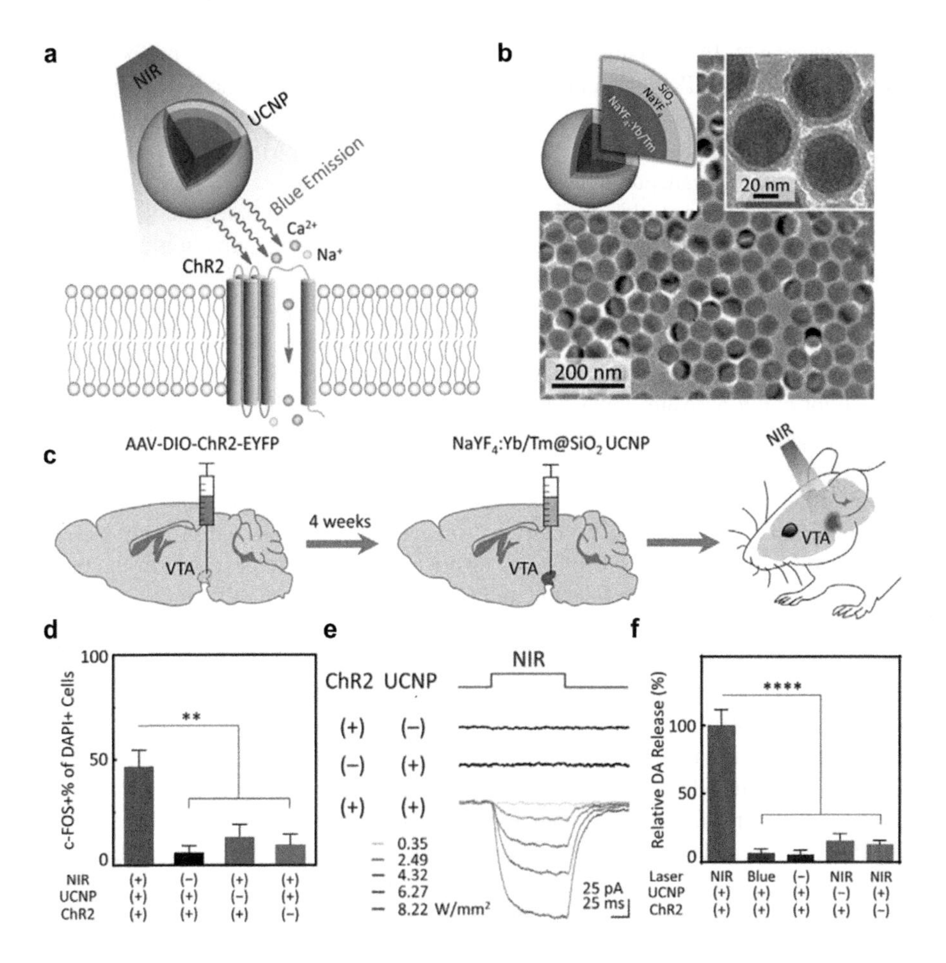

FIGURE 11.9 UCNP-mediated optogenetics in the deep brain. (a) Schematic illustration of UCNP-mediated NIR upconversion optogenetics for ion channel regulation. (b) Transmission electron microscope (TEM) image of NaYF4:Yb/Tm@SiO_2 nanoparticles. (c) Schematic illustration for the transcranial NIR stimulation of the VTA dopamine neurons in the mouse brain. (d) The proportion of c-Fos–positive cells in VTA under different conditions. (e) Patch-clamp recordings of neurons stimulated by NIR with different power intensities. (f) Transcraninal NIR stimulation caused dopamine release under five different conditions. (Reproduced with permission from Ref.[70]; Copyright (2018) American Association for the Advancement of Science (AAAS).)

possibilities for optogenetics modulation[74–76]. ML materials can emit light under various forms of mechanical stress, such as grinding, scraping, shaking, airflow, pressing, and ultrasonic pulse[76–78] (Figure 11.10a-c). This relatively easy method of light generation opens a new window of the application of mechanoluminescence, including new light source[79], anti-counterfeiting encryption[80], stress sensing[81], and wearable devices[82]. Typically, the emission wavelength of these ML materials covers

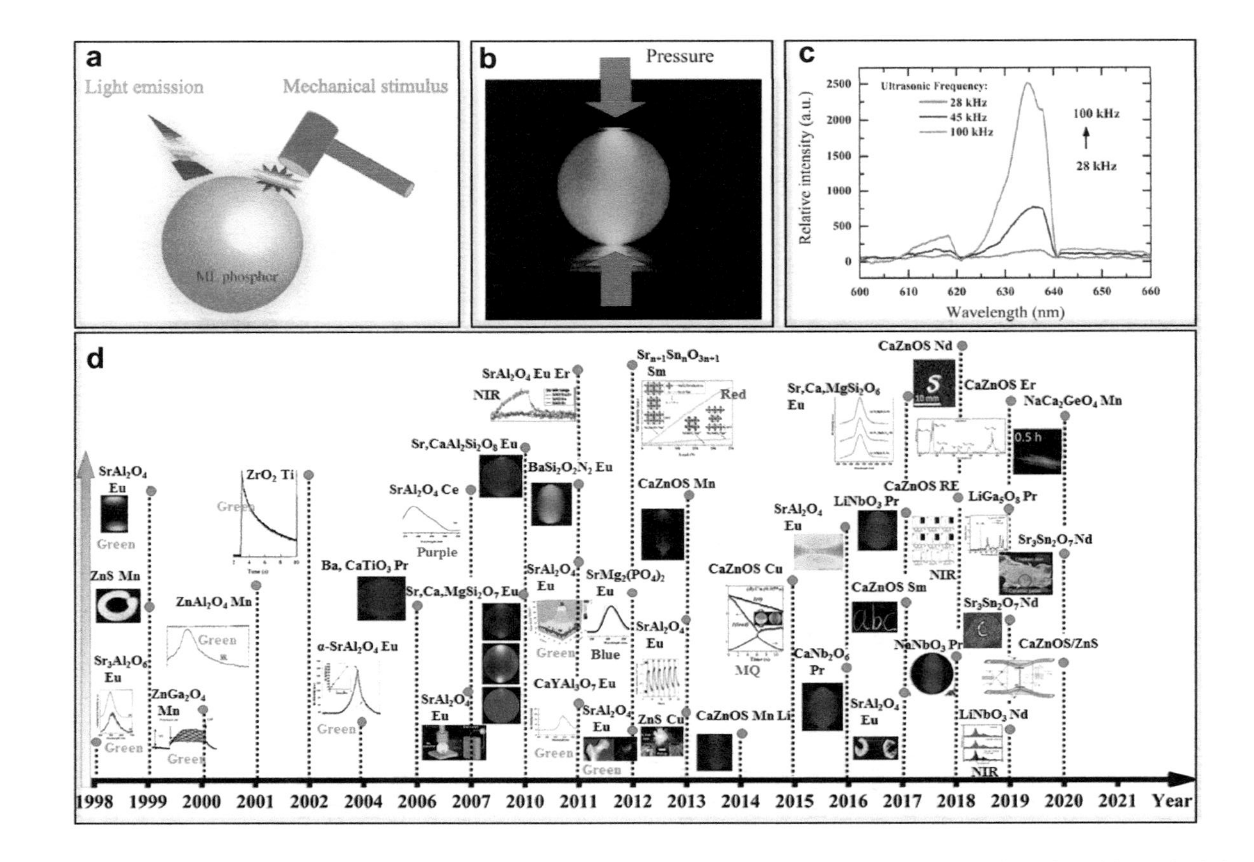

FIGURE 11.10 Schematic illustration of mechanoluminescence and development of ML materials. (a) Illustration of working principle of mechanoluminescent. (Reproduced with permission from Ref.[76]. Copyright (2015) John Wiley and Sons.) (b) Mechanoluminescence of the $SrAl_2O_4$:Eu (SAO) material. (Reproduced with permission from Ref.[77].) (c) The mechanoluminescent of the (1-x)$Pb(Mg_{1/3}Nb_{2/3})O_3$-x$PbTiO_3$ (PMN-PT) crystal with ultrasound treatment at the frequency of 28 kHz, 45 kHz and 100 kHz, respectively. (Reproduced from Ref.[78]; Licensed under CC BY 4.0.) (d) The development of mechanoluminescent materials from 1998–2021. (Reproduced with permission from Ref.[75]; Copyright (2021) Elsevier.)

the range from visible to near-infrared area[75] (Figure 11.10d), which makes the ML materials great candidates for applications in non-invasive optogenetics.

Recently, Hong et al. developed a new optogenetics modulation method based on mechanoluminescence[83]. When the ML nanoparticles pass through the intact scalp and skull, they are triggered by focused ultrasound (FUS) which can easily penetrate through the brain tissue, serving as the local light source for optogenetics. This method provides non-invasive optogenetics regulation in the brain without scalp incision, craniotomy, or brain implantation. As shown in Figure 11.11a, the ML nanoparticles enter the blood circulation through intravenous injection. During the circulation, they are energized by 400 nm ultraviolet light in the superficial blood vessels and activated by focused ultrasound. The light with 470 nm wavelength is repeatedly emitted in the mouse brain for optical stimulation. The intense luminescence of ML particles is capable of stimulating ChR2 expression neurons (Figure 11.11b and 11c). In this study, ZnS:Ag, Co@ZnS nanoparticles were injected into the blood vessels. Thy1-ChR2 YFP mice showed obvious hindlimb movement, which was synchronized with focused ultrasound stimulation. Under the same stimulation, the same animals did not show any hindlimb movement before injecting nanoparticles. Wild-type (WT) mice had no hindlimb movement regardless of the presence of ML nanoparticles in the blood circulation (Figure 11.11d). Overall, different from the traditional optogenetic method of optical fiber implantation, mechanoluminescent materials can be delivered intravenously and can be activated and silenced by ultrasound with temporal and spatial precision. This example also provides new possibilities for the *in vivo* application of other ML materials for optogenetics.

11.7 CONCLUSION

Advances in optogenetics have paved the way toward a better understanding of the nervous system, mainly due to its cell-specific, high spatiotemporal manipulation of neural activity. Despite its advantages, bottlenecks caused by conventional technologies prevent optogenetics from reaching its full potential. Tethered systems that use rigid implanted optical fibers restrict movement and cause tissue damage, hence are not suitable for long-term chronic studies of neural function. Furthermore, tethers make it challenging to perform social experiments involving two or more freely behaving animals in the same arena. Wireless battery-free optogenetic systems that employ implantable μLEDs as localized light sources present new opportunities for deciphering the neural dynamics of complex behaviors. Wireless power delivery mechanisms lead to fully implantable platforms that have a negligible impact on the natural behavior of animals. Wireless communication capabilities enable real-time control and addressability on a per-device basis, allowing the design of highly complicated stimulation schemes for free behavior and social interaction experiments. Multimodal devices that integrate optogenetics with other forms of stimulation and recording continue to emerge, enriching the possibilities for neural control, closed-loop therapeutics, and clinical monitoring. Yet, there remain challenges in realizing non-invasive optogenetic control systems. On this front, recent developments on upconversion and mechanoluminescent materials use light and ultrasound for non-invasive optogenetic stimulations, respectively. These non-invasive stimulation

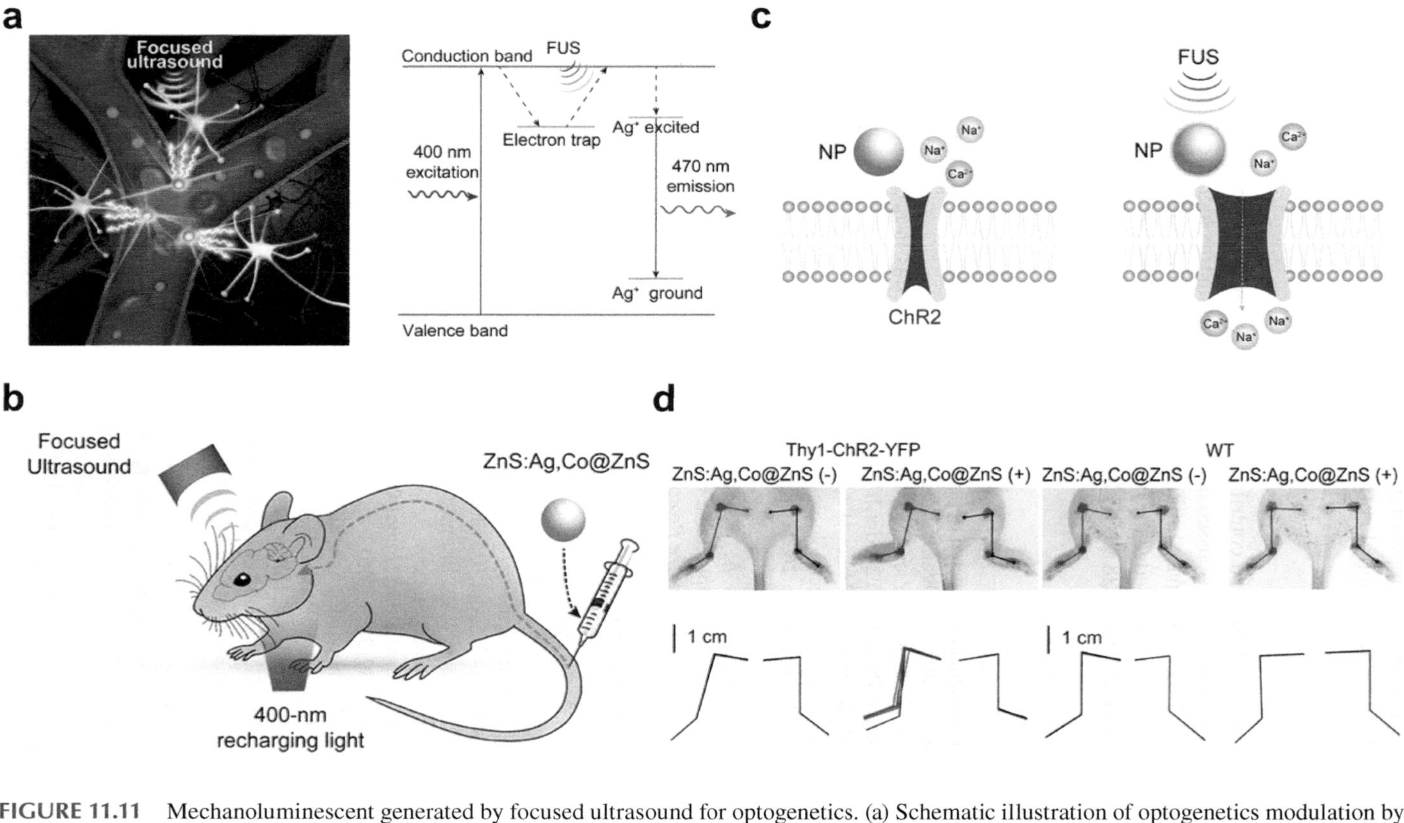

FIGURE 11.11 Mechanoluminescent generated by focused ultrasound for optogenetics. (a) Schematic illustration of optogenetics modulation by using focused ultrasound-activated light emission from ZnS:Ag, Co@ZnS nanoparticles. (b) Demonstration of *in vivo* sono-optogenetics stimulation through intact skull and scalp with intravenous injection of nanoparticles. (c) Focused ultrasound-assisted ChR2 channel modulation. (d) Kinematic diagrams of mouse hindlimbs before and after the injection of ZnS:Ag,Co@ZnS during sono-optogenetic stimulation. (Reproduced from Ref.[83]; Licensed under CC BY 4.0.)

techniques are increasingly appealing for achieving unprecedented spatiotemporal resolution and optogenetic control going forward.

ACKNOWLEDGMENT

Guangfu Wu and Vagif Abdulla contributed equally. The authors thank all researchers whose studies were cited and discussed in this book chapter and apologize to those whose work was not cited due to the space limitations. We thank Gavin Fennell for proofreading this book chapter. We acknowledge the funding support by the University of Connecticut start-up fund, NIH RF1NS118287, NIH R42MH116525, and NIH R61DA051489.

REFERENCES

1. Boyden, E. S., Zhang, F., Bamberg, E., Nagel, G. & Deisseroth, K. Millisecond-timescale, genetically targeted optical control of neural activity. *Nature Neuroscience* **8**, 1263–1268, doi:10.1038/nn1525 (2005).
2. Deisseroth, K. Optogenetics: 10 years of microbial opsins in neuroscience. *Nature Neuroscience* **18**, 1213–1225, doi:10.1038/nn.4091 (2015).
3. Deisseroth, K., Feng, G., Majewska, A. K., Miesenböck, G., Ting, A. & Schnitzer, M. J. Next-generation optical technologies for illuminating genetically targeted brain circuits. *Journal of Neuroscience* **26**, 10380–10386, doi:10.1523/JNEUROSCI.3863-06.2006 (2006).
4. Sparta, D. R., Stamatakis, A. M., Phillips, J. L., Hovelso, N., van Zessen, R. & Stuber, G. D. Construction of implantable optical fibers for long-term optogenetic manipulation of neural circuits. *Nature Protocols* **7**, 12–23, doi:10.1038/nprot.2011.413 (2011).
5. Jin, S., Li, J., Li, J., Lin, J. & Jiang, H. GaN microdisk light emitting diodes. *Applied Physics Letters* **76**, 631–633, doi:10.1063/1.125841 (2000).
6. Kim, T.-I., McCall, J. G., Jung, Y. H., Huang, X., Siuda, E. R., Li, Y., Song, J., Song, Y. M., Pao, H. A., Kim, R.-H., Lu, C., Lee, S. D., Song, I.-S., Shin, G., Al-Hasani, R., Kim, S., Tan, M. P., Huang, Y., Omenetto, F. G., Rogers, J. A. & Bruchas, M. R. Injectable, cellular-scale optoelectronics with applications for wireless optogenetics. *Science* **340**, 211–216, doi:10.1126/science.1232437 (2013).
7. Lee, S. T., Williams, P. A., Braine, C. E., Lin, D.-T., John, S. W. & Irazoqui, P. P. A miniature, fiber-coupled, wireless, deep-brain optogenetic stimulator. *IEEE Transactions on Neural Systems and Rehabilitation Engineering* **23**, 655–664, doi:10.1109/TNSRE.2015.2391282 (2015).
8. Kathe, C., Michoud, F., Schönle, P., Rowald, A., Brun, N., Ravier, J., Furfaro, I., Paggi, V., Kim, K., Soloukey, S., Asboth, L., Hutson, T. H., Jelescu, I., Philippides, A., Alwahab, N., Gandar, J., Huber, D., De Zeeuw, C. I., Barraud, Q., Huang, Q., Lacour, S. P. & Courtine, G. Wireless closed-loop optogenetics across the entire dorsoventral spinal cord in mice. *Nature Biotechnology* **40**, 198–208, doi:10.1038/s41587-021-01019-x (2022).
9. Lu, L., Gutruf, P., Xia, L., Bhatti, D. L., Wang, X., Vazquez-Guardado, A., Ning, X., Shen, X., Sang, T., Ma, R., Pakeltis, G., Sobczak, G., Zhang, H., Seo, D.-o., Xue, M., Yin, L., Chanda, D., Sheng, X., Bruchas, M. R. & Rogers, J. A. Wireless optoelectronic photometers for monitoring neuronal dynamics in the deep brain. *Proceedings of the National Academy of Sciences* **115**, E1374-E1383, doi:10.1073/pnas.1718721115 (2018).
10. Ausra, J., Wu, M., Zhang, X., Vázquez-Guardado, A., Skelton, P., Peralta, R., Avila, R., Murickan, T., Haney, C. R., Huang, Y., Rogers, J. A., Kozorovitskiy, Y. & Gutruf, P. Wireless, battery-free, subdermally implantable platforms for transcranial and

long-range optogenetics in freely moving animals. *Proceedings of the National Academy of Sciences* **118**, doi:10.1073/pnas.2025775118 (2021).

11. Yang, Y., Wu, M., Vázquez-Guardado, A., Wegener, A. J., Grajales-Reyes, J. G., Deng, Y., Wang, T., Avila, R., Moreno, J. A., Minkowicz, S., Dumrongprechachan, V., Lee, J., Zhang, S., Legaria, A. A., Ma, Y., Mehta, S., Franklin, D., Hartman, L., Bai, W., Han, M., Zhao, H., Lu, W., Yu, Y., Sheng, X., Banks, A., Yu, X., Donaldson, Z. R., Gereau IV, R. W., Good, C. H., Xie, Z., Huang, Y., Kozorovitskiy, Y. & Rogers, J. A. Wireless multilateral devices for optogenetic studies of individual and social behaviors. *Nature Neuroscience* **24**, 1035–1045, doi:10.1038/s41593–021–00849-x (2021).
12. Won, S. M., Cai, L., Gutruf, P. & Rogers, J. A. Wireless and battery-free technologies for neuroengineering. *Nature Biomedical Engineering*, 1–19, doi:10.1038/s41551-021-00683-3 (2021).
13. Montgomery, K. L., Yeh, A. J., Ho, J. S., Tsao, V., Mohan Iyer, S., Grosenick, L., Ferenczi, E. A., Tanabe, Y., Deisseroth, K., Delp, S. L. & Poon, A. S. Y. Wirelessly powered, fully internal optogenetics for brain, spinal and peripheral circuits in mice. *Nature Methods* **12**, 969–974, doi:10.1038/nmeth.3536 (2015).
14. Naik, N. & Perla, Y. Mechanical behaviour of acrylic under high strain rate tensile loading. *Polymer Testing* **27**, 504–512, doi:10.1016/j.polymertesting.2008.02.005 (2008).
15. Gefen, A., Gefen, N., Zhu, Q., Raghupathi, R. & Margulies, S. S. Age-dependent changes in material properties of the brain and braincase of the rat. *Journal of Neurotrauma* **20**, 1163–1177, doi:10.1089/089771503770802853 (2003).
16. Elkin, B. S., Azeloglu, E. U., Costa, K. D. & Morrison III, B. Mechanical heterogeneity of the rat hippocampus measured by atomic force microscope indentation. *Journal of Neurotrauma* **24**, 812–822, doi:10.1089/neu.2006.0169 (2007).
17. Kruse, S. A., Rose, G. H., Glaser, K. J., Manduca, A., Felmlee, J. P., Jack Jr, C. R. & Ehman, R. L. Magnetic resonance elastography of the brain. *Neuroimage* **39**, 231–237, doi:10.1016/j.neuroimage.2007.08.030 (2008).
18. Moore, S. W. & Sheetz, M. P. Biophysics of substrate interaction: Influence on neural motility, differentiation, and repair. *Developmental Neurobiology* **71**, 1090–1101, doi:10.1002/dneu.20947 (2011).
19. Park, S. I., Brenner, D. S., Shin, G., Morgan, C. D., Copits, B. A., Chung, H. U., Pullen, M. Y., Noh, K. N., Davidson, S., Oh, S. J., Yoon, J., Jang, K.-I., Samineni, V. K., Normal, M., Grajales-Reyes, J. G., Vogt, S. K., Sundaram, S. S., Wilson, K. M., Ha, J. S., Xu, R., Pan, T., Kim, T.-i., Huang, Y., Montana, M. C., Golden, J. P., Bruchas, M. R., Gereau IV, R. W. & Rogers, J. A. Soft, stretchable, fully implantable miniaturized optoelectronic systems for wireless optogenetics. *Nature Biotechnology* **33**, 1280–1286, doi:10.1038/nbt.3415 (2015).
20. Dobkin, D. *The RF in RFID: UHF RFID in practice* (Newnes, 2012).
21. Challis, L. Mechanisms for interaction between RF fields and biological tissue. *Bioelectromagnetics* **26**, S98–S106, doi:10.1002/bem.20119 (2005).
22. Michaelson S. M. Biological effects of radiofrequency radiation: concepts and criteria. *Health Physics* **61**, 3–14, doi:10.1097/00004032-199107000-00001 (1991).
23. Shin, G., Gomez, A. M., Al-Hasani, R., Jeong, Y. R., Kim, J., Xie, Z., Banks, A., Lee, S. M., Han, S. Y., Yoo, C. J., Lee, J.-L., Lee, S. H., Kurniawan, J., Tureb, J., Guo, Z., Yoon, J., Park, S. I., Bang, S. Y., Nam, Y., Walicki, M. C., Samineni, V. K., Mickle, A. D., Lee, K., Heo, S. Y., McCall, J. G., Pan, T., Wang, L., Feng, X., Kim, T.-i., Kim, J. K., Li, Y., Huang, Y., Gereau IV, R. W., Ha, J. S., Bruchas, M. R. & Rogers, J. A. Flexible near-field wireless optoelectronics as subdermal implants for broad applications in optogenetics. *Neuron* **93**, 509–521. e503, doi:10.1016/j.neuron.2016.12.031 (2017).
24. Gutruf, P., Krishnamurthi, V., Vázquez-Guardado, A., Xie, Z., Banks, A., Su, C.-J., Xu, Y., Haney, C. R., Waters, E. A., Kandela, I., Krishnan, S. R., Ray, T., Leshock, J.

P., Huang, Y., Chanda, D. & Rogers, J. A. Fully implantable optoelectronic systems for battery-free, multimodal operation in neuroscience research. *Nature Electronics* **1**, 652–660, doi:10.1038/s41928-018-0175-0 (2018).

25. Mickle, A. D., Won, S. M., Noh, K. N., Yoon, J., Meacham, K. W., Xue, Y., McIlvried, L. A., Copits, B. A., Samineni, V. K., Crawford, K. E., Kim, D. H., Srivastava, P., Kim, B. H., Min, S., Shiuan, Y., Yun, Y., Payne, M. A., Zhang, J., Jang, H., Li, Y., Lai, H. H., Huang, Y., Park, S. I., Gereau IV, R. W. & Rogers, J. A. A wireless closed-loop system for optogenetic peripheral neuromodulation. *Nature* **565**, 361–365, doi:10.1038/s41586-018-0823-6 (2019).
26. Weber, M. J., Bhat, A., Chang, T. C., Charthad, J. & Arbabian, A. *In 2016 IEEE topical conference on biomedical wireless technologies, networks, and sensing systems (BioWireleSS)*, 12–14 (IEEE).
27. Wirdatmadja, S. A., Balasubramaniam, S., Koucheryavy, Y. & Jornet, J. M. *In 2016 IEEE 18th international conference on e-health networking, applications and services (Healthcom)*, 1–6 (IEEE).
28. Wirdatmadja, S. A., Barros, M. T., Koucheryavy, Y., Jornet, J. M. & Balasubramaniam, S. Wireless optogenetic nanonetworks for brain stimulation: Device model and charging protocols. *IEEE Transactions on NanoBioscience* **16**, 859–872, doi:10.1109/TNB.2017.2781150 (2017).
29. Baumgartner, R. W. *Handbook on neurovascular ultrasound*. Vol. 21 (Karger Medical and Scientific Publishers, 2006).
30. ISO/IEC 15693–1:2018 www.iso.org/standard/70837.html (accessed Mar 22, 2022).
31. Yang, Y., Wu, M., Wegener, A. J., Vázquez-Guardado, A., Efimov, A. I., Lie, F., Wang, T., Ma, Y., Banks, A., Li, Z., Xie, Z., Huang, Y., Good, C. H., Kozorovitskiy, Y. & Rogers, J. A. Preparation and use of wireless reprogrammable multilateral optogenetic devices for behavioral neuroscience. *Nature Protocols*, **17**, 1073–1096, doi:10.1038/s41596-021-00672-5 (2022).
32. Kingsbury, L., Huang, S., Wang, J., Gu, K., Golshani, P., Wu, Y. E. & Hong, W. Correlated neural activity and encoding of behavior across brains of socially interacting animals. *Cell* **178**, 429–446. e416, doi:10.1016/j.cell.2019.05.022 (2019).
33. Kingsbury, L. & Hong, W. A multi-brain framework for social interaction. *Trends in Neurosciences* **43**, 651–666, doi:10.1016/j.tins.2020.06.008 (2020).
34. Bluetooth technology overview www.bluetooth.com/learn-about-bluetooth/tech-overview/ (accessed Mar 22, 2022).
35. Kim, C. Y., Ku, M. J., Qazi, R., Nam, H. J., Park, J. W., Nam, K. S., Oh, S., Kang, I., Jang, J.-H., Kim, W. Y., Kim, J.-H. & Jeong, J.-W. Soft subdermal implant capable of wireless battery charging and programmable controls for applications in optogenetics. *Nature Communications* **12**, 1–13, doi:10.1038/s41467–020–20803-y (2021).
36. Zhang, Y., Castro, D. C., Han, Y., Wu, Y., Guo, H., Weng, Z., Xue, Y., Ausra, J., Wang, X., Li, R., Wu, G., Vazquez-Guardado, A., Xie, Y., Xie, Z., Ostojich, D., Peng, D., Sun, R., Wang, B., Yu, Y., Leshock, J. P., Qu, S., Su, C.-J., Shen, W., Hang, T., Banks, A., Huang, Y., Radulovic, J., Gutruf, P., Bruchas, M. R. & Rogers, J. A. Battery-free, lightweight, injectable microsystem for in vivo wireless pharmacology and optogenetics. *Proceedings of the National Academy of Sciences* **116**, 21427–21437, doi:10.1073/pnas.1909850116 (2019).
37. Zhang, Y., Mickle, A. D., Gutruf, P., McIlvried, L. A., Guo, H., Wu, Y., Golden, J. P., Xue, Y., Grajales-Reyes, J. G., Wang, X., Krishnan, S. R., Xie, Y., Peng, D., Su, C.-J., Zhang, F., Reeder, J. T., Vogt, S. K., Huang, Y., Rogers, J. A. & Gereau IV, R. W. Battery-free, fully implantable optofluidic cuff system for wireless optogenetic and pharmacological neuromodulation of peripheral nerves. *Science Advances* **5**, eaaw5296, doi:10.1126/sciadv.aaw5296 (2019).
38. Wu, G., Heck, I., Zhang, N., Phaup, G., Zhang, X., Wu, Y., Stalla, D. E., Weng, Z., Sun, H., Li, H., Zhang, Z., Ding, S., Li, D.-P. & Zhang, Y. Wireless, battery-free push-pull

microsystem for membrane-free neurochemical sampling in freely moving animals. *Science Advances* **8**, eabn2277, doi:10.1126/sciadv.abn2277 (2022).

39. Park, S., Guo, Y., Jia, X., Choe, H. K., Grena, B., Kang, J., Park, J., Lu, C., Canales, A., Chen, R., Yim, Y. S., Choi, G. B., Fink, Y. & Anikeeva, P. One-step optogenetics with multifunctional flexible polymer fibers. *Nature Neuroscience* **20**, 612–619, doi:10.1038/nn.4510 (2017).
40. Burton, A., Obaid, S. N., Vázquez-Guardado, A., Schmit, M. B., Stuart, T., Cai, L., Chen, Z., Kandela, I., Haney, C. R., Waters, E. A., Cai, H., Rogers, J. A., Lu, L. & Gutruf, P. Wireless, battery-free subdermally implantable photometry systems for chronic recording of neural dynamics. *Proceedings of the National Academy of Sciences* **117**, 2835–2845, doi:10.1073/pnas.1920073117 (2020).
41. Zhang, H., Gutruf, P., Meacham, K., Montana, M. C., Zhao, X., Chiarelli, A. M., Vázquez-Guardado, A., Norris, A., Lu, L., Guo, Q., Xu, C., Wu, Y., Zhao, H., Ning, X., Bai, W., Kandela, I., Haney, C. R., Chanda, D., Gereau IV, R. W. & Rogers, J. A. Wireless, battery-free optoelectronic systems as subdermal implants for local tissue oximetry. *Science Advances* **5**, eaaw0873, doi:10.1126/sciadv.aaw0873 (2019).
42. Nakai, J., Ohkura, M. & Imoto, K. A high signal-to-noise Ca2+ probe composed of a single green fluorescent protein. *Nature Biotechnology* **19**, 137–141, doi:10.1038/84397 (2001).
43. Wen, S., Zhou, J., Zheng, K., Bednarkiewicz, A., Liu, X. & Jin, D. Advances in highly doped upconversion nanoparticles. *Nature Communications* **9**, 2415, doi:10.1038/s41467-018-04813-5 (2018).
44. Chen, B. & Wang, F. Emerging frontiers of upconversion nanoparticles. *Trends in Chemistry* **2**, 427–439, doi:10.1016/j.trechm.2020.01.008 (2020).
45. Liu, X., Yan, C. H. & Capobianco, J. A. Photon upconversion nanomaterials. *Chemical Society Reviews* **44**, 1299–1301, doi:10.1039/c5cs90009c (2015).
46. Sun, Y., Feng, W., Yang, P., Huang, C. & Li, F. The biosafety of lanthanide upconversion nanomaterials. *Chemical Society Reviews* **44**, 1509–1525, doi:10.1039/c4cs00175c (2015).
47. Feng, W., Zhu, X. & Li, F. Recent advances in the optimization and functionalization of upconversion nanomaterials for in vivo bioapplications. *NPG Asia Materials* **5**, e75–e75, doi:10.1038/am.2013.63 (2013).
48. Zhou, B., Shi, B., Jin, D. & Liu, X. Controlling upconversion nanocrystals for emerging applications. *Nature Nanotechnology* **10**, 924–936, doi:10.1038/nnano.2015.251 (2015).
49. Chen, X., Peng, D., Ju, Q. & Wang, F. Photon upconversion in core-shell nanoparticles. *Chemical Society Reviews* **44**, 1318–1330, doi:10.1039/c4cs00151f (2015).
50. Sun, L., Wei, R., Feng, J. & Zhang, H. Tailored lanthanide-doped upconversion nanoparticles and their promising bioapplication prospects. *Coordination Chemistry Reviews* **364**, 10–32, doi:10.1016/j.ccr.2018.03.007 (2018).
51. Zhong, Y., Tian, G., Gu, Z., Yang, Y., Gu, L., Zhao, Y., Ma, Y. & Yao, J. Elimination of photon quenching by a transition layer to fabricate a quenching-shield sandwich structure for 800 nm excited upconversion luminescence of Nd3+-sensitized nanoparticles. *Advanced Materials* **26**, 2831–2837, doi:10.1002/adma.201304903 (2014).
52. Zheng, K., Loh, K. Y., Wang, Y., Chen, Q., Fan, J., Jung, T., Nam, S. H., Suh, Y. D. & Liu, X. Recent advances in upconversion nanocrystals: Expanding the kaleidoscopic toolbox for emerging applications. *Nano Today* **29**, 100797, doi:10.1016/j.nantod.2019.100797 (2019).
53. Wang, Z., Hu, M., Ai, X., Zhang, Z. & Xing, B. Near-Infrared Manipulation of Membrane Ion Channels via Upconversion Optogenetics. *Advanced Biosystems* **3**, e1800233, doi:10.1002/adbi.201800233 (2019).
54. Himmelstoß, S. F. & Hirsch, T. A critical comparison of lanthanide based upconversion nanoparticles to fluorescent proteins, semiconductor quantum dots, and carbon

dots for use in optical sensing and imaging. *Methods and Applications in Fluorescence* **7**, 022002, doi:10.1088/2050–6120/ab0bfa (2019).
55. Gu, Z., Yan, L., Tian, G., Li, S., Chai, Z. & Zhao, Y. Recent advances in design and fabrication of upconversion nanoparticles and their safe theranostic applications. *Advanced Materials* **25**, 3758–3779, doi:10.1002/adma.201301197 (2013).
56. Chen, G., Qiu, H., Prasad, P. N. & Chen, X. Upconversion nanoparticles: design, nanochemistry, and applications in theranostics. *Chemical Reviews* **114**, 5161–5214, doi:10.1021/cr400425h (2014).
57. Wu, S. & Butt, H. J. Near-infrared-sensitive materials based on upconverting nanoparticles. *Advanced Materials* **28**, 1208–1226, doi:10.1002/adma.201502843 (2016).
58. Deisseroth, K. & Anikeeva, P. Upconversion of light for use in optogenetic methods. *United States Patent, PCT/US11/59287* (2016).
59. Qin, X., Xu, J., Wu, Y. & Liu, X. Energy-transfer editing in lanthanide-activated upconversion nanocrystals: A toolbox for emerging applications. *ACS Central Science* **5**, 29–42, doi:10.1021/acscentsci.8b00827 (2019).
60. Hososhima, S., Yuasa, H., Ishizuka, T., Hoque, M. R., Yamashita, T., Yamanaka, A., Sugano, E., Tomita, H. & Yawo, H. Near-infrared (NIR) up-conversion optogenetics. *Scientific Reports* **5**, 16533, doi:10.1038/srep16533 (2015).
61. Shah, S., Liu, J. J., Pasquale, N., Lai, J., McGowan, H., Pang, Z. P. & Lee, K. B. Hybrid upconversion nanomaterials for optogenetic neuronal control. *Nanoscale* **7**, 16571–16577, doi:10.1039/c5nr03411f (2015).
62. Wu, X., Zhang, Y., Takle, K., Bilsel, O., Li, Z., Lee, H., Zhang, Z., Li, D., Fan, W., Duan, C., Chan, E. M., Lois, C., Xiang, Y. & Han, G. Dye-sensitized core/active shell upconversion nanoparticles for optogenetics and bioimaging applications. *ACS Nano* **10**, 1060–1066, doi:10.1021/acsnano.5b06383 (2016).
63. Bansal, A., Liu, H., Jayakumar, M. K., Andersson-Engels, S. & Zhang, Y. Quasi-continuous wave near-infrared excitation of upconversion nanoparticles for optogenetic manipulation of C. elegans. *Small* **12**, 1732–1743, doi:10.1002/smll.201503792 (2016).
64. Ao, Y., Zeng, K., Yu, B., Miao, Y., Hung, W., Yu, Z., Xue, Y., Tan, T. T. Y., Xu, T., Zhen, M., Yang, X., Zhang, Y. & Gao, S. An upconversion nanoparticle enables near infrared-optogenetic manipulation of the caenorhabditis elegans motor circuit. *ACS Nano* **13**, 3373–3386, doi:10.1021/acsnano.8b09270 (2019).
65. Zhang, Y., Zhang, W., Zeng, K., Ao, Y., Wang, M., Yu, Z., Qi, F., Yu, W., Mao, H., Tao, L., Zhang, C., Tan, T. T. Y., Yang, X., Pu, K. & Gao, S. Upconversion nanoparticles-based multiplex protein activation to neuron ablation for locomotion regulation. *Small* **16**, e1906797, doi:10.1002/smll.201906797 (2020).
66. Ai, X., Lyu, L., Zhang, Y., Tang, Y., Mu, J., Liu, F., Zhou, Y., Zuo, Z., Liu, G. & Xing, B. Remote regulation of membrane channel activity by site-specific localization of lanthanide-doped upconversion nanocrystals. *Angewandte Chemie International Edition* **56**, 3031–3035, doi:10.1002/anie.201612142 (2017).
67. Lin, X., Wang, Y., Chen, X., Yang, R., Wang, Z., Feng, J., Wang, H., Lai, K. W., He, J., Wang, F. & Shi, P. Multiplexed optogenetic stimulation of neurons with spectrum-selective upconversion nanoparticles. *Advanced Healthcare Materials* **6**, 1700446, doi:10.1002/adhm.201700446 (2017).
68. Wang, Y., Lin, X., Chen, X., Chen, X., Xu, Z., Zhang, W., Liao, Q., Duan, X., Wang, X., Liu, M., Wang, F., He, J. & Shi, P. Tetherless near-infrared control of brain activity in behaving animals using fully implantable upconversion microdevices. *Biomaterials* **142**, 136–148, doi:10.1016/j.biomaterials.2017.07.017 (2017).
69. Lin, X., Chen, X., Zhang, W., Sun, T., Fang, P., Liao, Q., Chen, X., He, J., Liu, M., Wang, F. & Shi, P. Core—shell—shell upconversion nanoparticles with enhanced

emission for wireless optogenetic inhibition. *Nano Letters* **18**, 948–956, doi:10.1021/acs.nanolett.7b04339 (2018).

70. Chen, S., Weitemier, A. Z., Zeng, X., He, L., Wang, X., Tao, Y., Huang, A. J. Y., Hashimotodani, Y., Kano, M., Iwasaki, H., Parajuli, L. K., Okabe, S., Teh, D. B. L., All, A. H., Tsutsui-Kimura, I., Tanaka, K. F., Liu, X. & McHugh, T. J. Near-infrared deep brain stimulation via upconversion nanoparticle-mediated optogenetics. *Science* **359**, 679–684, doi:10.1126/science.aaq1144 (2018).
71. Nguyen, N. T., Huang, K., Zeng, H., Jing, J., Wang, R., Fang, S., Chen, J., Liu, X., Huang, Z., You, M. J., Rao, A., Huang, Y., Han, G. & Zhou, Y. Nano-optogenetic engineering of CAR T cells for precision immunotherapy with enhanced safety. *Nature Nanotechnology* **16**, 1424–1434, doi:10.1038/s41565-021-00982-5 (2021).
72. Tan, P., He, L., Huang, Y. & Zhou, Y. Optophysiology: Illuminating cell physiology with optogenetics. *Physiological Reviews* **102**, 1263–1325, doi:10.1152/physrev.00021.2021 (2022).
73. He, L., Tan, P., Zhu, L., Huang, K., Nguyen, N. T., Wang, R., Guo, L., Li, L., Yang, Y., Huang, Z., Huang, Y., Han, G., Wang, J. & Zhou, Y. Circularly permuted LOV2 as a modular photoswitch for optogenetic engineering. *Nature Chemical Biology* **17**, 915–923, doi:10.1038/s41589-021-00792-9 (2021).
74. Chen, B., Zhang, X. & Wang, F. Expanding the toolbox of inorganic mechanoluminescence materials. *Accounts of Materials Research* **2**, 364–373, doi:10.1021/accountsmr.1c00041 (2021).
75. Xiong, P., Peng, M. & Yang, Z. Near-infrared mechanoluminescence crystals: A review. *Iscience* **24**, 101944 (2021).
76. Peng, D., Chen, B. & Wang, F. Recent advances in doped mechanoluminescent phosphors. *ChemPlusChem* **80**, 1209–1215, doi:10.1002/cplu.201500185 (2015).
77. Bu, N., Ueno, N., Xu, C. & Fukuda, O. Measurement of weak light emitted from mechanoluminescence materials using Si photodiode and light concentrator. *SENSORS*, 1528–1532, doi: 10.1109/ICSENS.2009.5398472 (2009).
78. Jia, Y., Tian, X., Wu, Z., Tian, X., Zhou, J., Fang, Y. & Zhu, C. Novel mechano-luminescent sensors based on piezoelectric/electroluminescent composites. *Sensors* **11**, 3962–3969, doi:10.3390/s110403962 (2011).
79. Wang, X., Peng, D., Huang, B., Pan, C. & Wang, Z. L. Piezophotonic effect based on mechanoluminescent materials for advanced flexible optoelectronic applications. *Nano Energy* **55**, 389–400, doi:10.1016/j.nanoen.2018.11.014 (2019).
80. Wang, H., Chen, X., Tian, Z., Jiang, Z., Yu, W., Ding, W. & Su, L. Efficient color manipulation of zinc sulfide-based mechanoluminescent elastomers for visualized sensing and anti-counterfeiting. *Journal of Luminescence* **228**, 117590, doi:10.1016/j.jlumin.2020.117590 (2020).
81. Zhuang, Y. & Xie, R. J. Mechanoluminescence rebrightening the prospects of stress sensing: A review. *Advanced Materials* **33**, e2005925, doi:10.1002/adma.202005925 (2021).
82. Wang, C., Yu, Y., Yuan, Y., Ren, C., Liao, Q., Wang, J., Chai, Z., Li, Q. & Li, Z. Heartbeat-sensing mechanoluminescent device based on a quantitative relationship between pressure and emissive intensity. *Matter* **2**, 181–193, doi:10.1016/j.matt.2019.10.002 (2020).
83. Wu, X., Zhu, X., Chong, P., Liu, J., Andre, L. N., Ong, K. S., Brinson, K., Jr., Mahdi, A. I., Li, J., Fenno, L. E., Wang, H. & Hong, G. Sono-optogenetics facilitated by a circulation-delivered rechargeable light source for minimally invasive optogenetics. *Proceedings of the National Academy of Sciences*, **116** (52) 26332–26342, doi:10.1073/pnas.1914387116 (2019).

Index

Note: numbers in **bold** indicate a table. Numbers in *italics* indicate a figure.

μCs (microcontrollers), 158, 159, 161, 162
μLEDs (microscale light-emitting diodes), 152, 154, *155*, 156, *157*, *162*, 163, 171
μPDs (microscale photodetectors), 162–163

A

AAV, *see* adeno-associated virus
accidental cell death (ACD), 88
adeno-associated virus (AAV), 127–129
adeno-associated virus (AAV) delivery
  in vivo DNA recombination with FISC system using, 143–145
adeno-associated virus (AAV)-mediated FISC system, 137, 145–146
Alzheimer's disease (AD), 37, 112
anti-BrdU antibody, 81
anti-counterfeiting encryption, 168
anti-Stokes luminescence, 163
anti-tumor immune response, 94
apoptosis, 37, **87**, 90–92
  Ca^{2+}-mediated, 165
  optogenetic control of, 88–89
Arabidopsis thaliana, 38, 87, 100
Arabidopsis thaliana cryptochrome 2 (AtCRY2), 4, 10, 26, 38, 72, 112, 124
autophagy-dependent cell death, 88
axon regeneration, 56, 62–68
Avena sativa (As), 2, 26, 27, 102, 116
  see also, light-oxygen-voltage (LOV) domains

B

Bluetooth (BT), 158, 160
Bluetooth Low Energy (BLE), 160
brain, *see* mouse brain
brain and beyond
  optogenetics and, 123–132
brain-derived neurotrophic factor (BDNF), *49*, 51
  Tropomyosin receptor kinase B (TrkB) and, 37, 38, 46, 48
budding yeast, 100, 103
  light-sensitive protein synthesis in, 101–103

C

C4da, *see* class IV da (C4da) neurons
Ca^{2+} channels, optogenetic engineering of
  cell-based screening with Ca^{2+} imaging, 28–30
  light activated, 25–33
  LOCA library, generation of 30–31
  LOV2-ORAI1 hybrid constructs, 27–29
  NFAT-dependent luciferase expression, 32–33
  NFAT translocation-based high-content imaging, 32
  ORAI1 expression vector, 27
Ca^{2+} release-activated Ca^{2+} (CRAC) channel, 2, 15, 17, 25
  CRAC channel-based optogenetics for versatile control of Ca^{2+} signaling, 5–10
Caenorhabditis elegans (*C. elegans*, nematode worm), 100, 112, 165
calcium
  genetically encoded calcium indicators (GECIs), 163
  intracellular, 42
  store-operated calcium entry (SOCE), 2
  two-component influx pathway, 1–17
  see also, Ca^{2+} channel; Ca^{2+} release-activated Ca^{2+} (CRAC) channel
caspase
  caspase-1, **87**, 91, 92
  caspase-3, 88, *89*, 91, 165
  caspase-4, **87**
  caspase-5, **87**, 91, *92*
  caspase-8, **87**, 88, 90, 91
  caspase-9, **87**, 88
  human caspase-4, 92
  murine caspase-11, 91
  optoCaspase-8/9, 88, 89, 92
  RIPK1-FADD-caspase-8 complex, 91
caspase recruitment domain (CARD), 88
  CARD/DED-deficient optoCaspase-8/9, 89
CDC20, 104, **105**
cell death manipulation, *see* programmed cell death manipulation
cell transfection
  HSC, 79–80
channelrhodopsins, 56, 111
  channelrhodopsin 2 (ChR2), 25, 26, 85, 165, 167
  thy1-ChR2 YFP, 170, *171*
Chlamydomonas reinhardtii, vii

class IV da (C4da) neurons, 60, 62–67
COH2, 140
CRAC, *see* Ca^{2+} release-activated Ca^{2+} (CRAC) channel
CreC, *see* C-terminal Cre fragment
Cre/*loxP* (Cre-loxP) 124, 136, 147
  FRL-induced Split system, 137
CreN59, 140
CRISPR, 105
Cryptochrome Circadian Regulator 2 (CRY2)
  CRY2, 4–7, 11
  CRY2-CIB1, 4, **86**, 87, 88
  CRY2-PHR, 4, **87**
  CRY2-STIM1ct, 12
  mCh-CRY2-STIM1233–448, 11
  mCh-CRY2-STIM1443–685, 7, *13*
  CRY2-TIM1ct, 14
cryptochrome-interacting basic-helix-loop-helix (CIB1), 56, 87
cryptochromes, 4, 85
C-terminal Cre fragment (CreC), 136, 140
  CreC60, 137, 138, 140, 141, 143

D

Danio rerio (zebrafish), 100, 112, 112, 165
dimerization, 1–4
  CIBN-CRY2, 72
  CIBN heterodimerization, 57
  controlling, 38–39
  CRY2-CIB1 heterodimerization, 86, 87, 88
  CRY-based, *115*
  EGFR, 114
  light-induced heterodimerization, 56
  LOV, 112, 116
  protein-protein heterodimerization, 17
  TrkA homodimerization, 73
  TrkB, 38
DNA ligation method, 39
DNA recombination
  FISC-mediated in mice, *145*, 146–147
  performance of FISC system in mammalian cells, 140–142
  in tdTomato transgenic mice, 143–146
dopamine, 167
Dronpa (DRONPA), 112, 114
  DRONPA-based caging, *118*
  DRONPA-derived photoswitchable kinases, 117
  engineered, 4
Drosophila melanogaster (fruit fly), 15, *16*, 60, 112, 119
  neuron injury model, 62–63
Dulbecco's phosphate buffered saline (DPBS), 73

E

entotic cell death, 88
Erg9, 105–107
extracellular signal-regulated kinase (ERK)
  anti-ERK, 50
  ERK1/2, 50, 76, 77, 81
  ERK-EGFP, 57
  ERK-GFP, 57, *58*
  ERK-KTR-GFP, 42, *43*
  pathways, 73
  pERK, 50–51, 57, *59*, 61–62, *63*
  psMEK action on, 117
  Raf/MEK/ERK activation, *75, 77*
  Raf/MEK/ERK signaling, 60, 65, 67, 72
  see also, mitogen-activated protein kinase//extracellular signal-regulated kinase (MAPK/ERK)

F

far red light (FRL)-dependent DNA recombination, 142
far red light (FRL)-induced Cre-*loxP* system (FISC system), 136–147
  DNA recombination performance in mammalian cells, 140–142
  FISC-controlled transgene expression in mammalian cells, *139*
  FISC-mediated DNA recombination in mice, 146
  FISC-mediated transgene expression in mammalian cells, *142*
  in vivo DNA recombination using AAV delivery, 143–145
  in vivo DNA recombination using hydrodynamic injection delivery, 143
  schematic representation of, *138*
far red light (FRL)-responsive chimeric promoters, 137–138, 141
ferroptosis, 88
FISC system, *see* far red light (FRL)-induced Cre-*loxP* system
flavin adenine dinucleotide (FAD) photoreduction, 87
flavin mononucleotide (FMN), 85
  cofactor, 27

G

GAL4, 60, 64
gasdermin family proteins, 91, 93
GDH1
  optogenetic control of, 105–106
genetically encoded calcium indicators (GECIs), 28, 29, 32, 163

guanylate triphosphate (GTP), 137, *138*
GTPase RAS, 72
GTPase Rit and Rin, 14
GTPase signaling, 27

H

hair-follicle-derived stem cells (HSC), 73, 79–82
HEK293 cells, 9, 56, *59*, *141*, 165
HEK293T cells, 32, 74, 127, *128*
HeLa cell lines, *8*, *11*, *12*, *16*, 28, 29, *33*, 125
NFAT, *31*, 32
hippocampus, 167
see also, mouse hippocampal dentate gyrus
Hippo signaling, 117
HSC, *see* hair-follicle-derived stem cells
human embryonic kidney (HEK) cell lines
HEK293, *9*, 28, 56, *59*, *141*, *142*, 165
HEK293T, 32, 74, 127, 146

I

immunogenic cell death, 88
inhibitors, 51, 112, 114–116, 119
protease, 75
integrated circuit cards, 158
integrated circuits (ICs), 152
inter-integrated circuits (I2C), 159
iodixanol buffers, 128–129
iodixanol gradient solution, **146**

L

lead zirconate titanate (PZT), 158
lentiviral transfer plasmids, *75*
lentivirus, 127, 129
lentivirus plasmid construction, 74
lentivirus preparation and neural progenitor, 74, *76*
light activatable Ca^{2+} channel, high throughput engineering of, 25–33
see also, Ca^{2+} channels, optogenetic engineering of
light-emitting diodes (LEDs)
far-red, *142*
microscale (μLEDs), 152, 154, *155*, 156, *157*, *162*, 163, 171
light-oxygen-voltage (LOV) domains, 2, 100
AuLOV, 38–41, 48, 51, 73, 79
Avena sativa (AsLOV2), 2, *3*, **102**, 102–103, 112
circularly permutated (cpLOV2), 2, *3*, 7, 8, **86–87**, 90, 91, 92, 93
different psTF variants related to, **102**
Jα-helix, 2, *86*
LOV2, 4–10
LOV2-ORAI1 hybrid constructs, 26, 27–28, 29, 30, 33
LOV2-SOAR, 7, 9, 10
LOV2-STIM1ct, 14
LOV2 Trap and Release of Protein (LOVTRAP), **86**, 93, 102, 112
Lyn-iTrkB-AuLOV-GFP, 41
Lyn-iTrkB-AuLOV-mCh, 39, 51
Lyn-TrkAICD-AuLOV-GFP, 41, 79
photoactivation window of, 94
tetR-*As*LOV2, 104, 105, 107
Vaucheria frigida, LOV domain of (VfAuLOV), 38
LiMETER2, 14, *15*
LipoStem Reagent, 79
LOV domains, *see* light-oxygen-voltage (LOV) domains
LRP6 receptor, 114
luciferase
Bright-Glo Luciferase Assay System, 33
NFAT-dependent luciferase expression, 32–33
LUXEON rebel LED, 60
Lyn, 39, 124
Lyn-AuLOV-TrkA(ICD)-GFP fusion protein, 73
Lyn-cytosolic Fas domain-Cry2PHR-EGFP, 125
Lyn-iTrkB-AuLOV-GFP, 41
Lyn-iTrkB-AuLOV-mCh, 39, 51
Lyn-TrkAICD-AuLOV-GFP, 41, 79
lysosome-dependent cell death, 88

M

mechanoluminescence (ML), 151, 167–171
mice
FISC-mediated DNA recombination in, *145*, 146
implanted, 154, *155*
monSTIM1, 124, 129
mPFC stimulation of, 160
social preference/interaction with synchronized, *159*
striatal dopamine released in, 167
tdTomato transgenic, 143, *144*
wild-type, 137
microbes, 38
microcontrollers (μCs), 158, 159, 161, 162
microfabrication, 152
microfluidics, 161, *162*
microorganisms, 100, 103
microscale light-emitting diodes (μLEDs), 152, 154, *155*, 156, *157*, *162*, 163, 171

microscale photodetectors (μPDs), 162–163
microscopy
Laser Scanning Microscopy (LSM), 119
optigenetics and, 117–119
microtubules (MT), 5, 13
recruitment of, 14
mitogen-activated protein kinase (MAPK) pathway, 112
Ras/MAPK 116
mitogen-activated protein kinase//extracellular signal-regulated kinase (MAPK/ERK), 41, 114
activation of, 42, *43*
TrkB downstream, 51
mouse, *see* mice
mouse bladder, 161
mouse brain, 124
fiber optic cannula implantation in, 129–131
hippocampal dentate gyrus (DG), 125
optical stimulation of, 170, *171*
probes implanted in, 165
transcranial NIR stimulation of VTA in, 168
upconversion optigenetics, development of *166*
viral vectors for, 126
virus injection into, 129–130
mouse neural progenitors, 73–76
mouse organs, 137

N

near-field communication (NFC), 156, 158, 159, 161
necrosis, 89
necroptosis, **87**, 88–91
nematode worm, *see Caenorhabditis elegans* (*C. elegans*)
NETotgic cell death, 88
neural activity, 119, 161, 165, 171
neural circuits, 151
neural differentiation, 46
neural progenitor infection, 74
neural progenitors, 72, 74
differentiation of, *77*
see also mouse neural progenitors
neural regeneration, 73, 82
neural repair
spatiotemporal modulation of, 55–67
see also, optoAKT; optogenetic systems; optoRaf
neural stem cell differentiation
optogenetic control of, 71–82
neural tissue, 154
neuritogenesis, 60
neurogenetic and signal transduction, optogenetics as tool to study, 111–119
neuropharmacology, 161, *162*
neurotrophin signaling pathway, 55, 68
neurotrophins, 56, 64, 73
p75 neurotrophin receptor (p75NTR), 37
PC12 cells exposed to, 46
NFAT, *see* nuclear factor of activated-T cells
Nomenclature Committee on Cell Death (NCCD) guidelines, 88
non-opsin-based light-sensitive proteins, 87
non-opsin optogenetic tools, 93, 119, 130, 132
non-opsin photoreceptors, **86**
non-opsin photosensitive proteins, 112
nuclear factor of activated-T cells (NFAT), 2, 25–26
HeLa, *31*
NFAT-dependent gene expression, 7
NFAT-dependent luciferase expression, 32
NFAT translocation-based high-content imaging, 32

O

oligomerization, 1–6, *8*, 26
BAK and BAX, 88
chemical inducible, 14
CRY2, 91, 112, 114, 124
CRY2-BAX, 89
CRY2-mCherry, 113, *115*
CRY2 mutants with increased tendency towards, 51
homo-, 38, 86, 87
photo-inducible dimerization and, *3*, 4
protein self-oligomerization, 10–12, 17
RIPK1, 89–90
oligonucleotide sequences, 104, **105**
OptoAKT
activation *in vivo* of, 60–62
activation kinetics of, *61*
axon regeneration capacity and, 65, *66*, *67*
biochemical and cellular responses induced by activation of, *59*, 60
design of, 56
fly sensory neuron injury model and, 62–64
generation of, 60
inactivation of, 62, *63*
reversible optogenetic stimulation of, *58*
scheme for, *57*
opto-cytTrkB(E281A), 124, 129
opto-FAS, 125, 129
optofluidics, 161, *162*
optogenetic photometry, 162
optogenetics
brain and beyond, 123–132
definition of, 1, 85, 151

ion signaling pathways controlled by, 85
microscopy and, 117–119
monSTIM1, 124
neurogenetic and signal transduction studied via, 111–119
non-invasive, 163–167
opto-cytTrkB(E281A), 124, 129
opto-FAS, 125, 129
photoactivatable Flp Recombinase (PA-Flp), 124
protein-aggregated-related diseases studied via, 112
proteins of interest (POI) and, 2–3, 38, 85, *86*
signalling pathways studied via, 114
sono-, 167–172
synergistic optogenetic multistep control of protein levels, 103–105
see also, channelrhodopsins; cryptochrome; LOV domains; *Saccharomyces cerevisiae*; wireless and non-invasive optogenetics
optogenetic regulation of protein stability in *Saccharomyces cerevisiae, see Saccharomyces cerevisiae*
optogenetic stimulation, 62–66
optogenetic systems
based on caging, 117
CRY2-dependent, 114; *see also* Cryptochrome Circadian Regulator 2 (CRY2)
in flies, activation of, 60–62
in vitro, 56–60
optogenetic toolbox
light-sensitive allosteric switches, 2–4
overview of, 2
photo-inducible dimerization or oligomerization systems, 4
photo-inducible dissociation system, 4
optogenetic tools, 103
OptoRaf
activation *in vivo* of, 60–62
activation kinetics of, *61*
axon regeneration capacity and, 65, *66, 67*
biochemical and cellular responses induced by activation of, *59*, 60
design of, 56
fly sensory neuron injury model and, 62–64
generation of, 60
inactivation of, 62, *63*
reversible optogenetic stimulation of, *58*
scheme for, *57*
see also, OptoRaf1
OptoRaf1
optogenetic control of mouse neural progenitors via, 73– 77, *77–78*, 82
Opto-SOS, 116
OptoTrkA, 71, 73
optogenetic control of hair follicle-derived stem cell differentiation via, 79–82
OptoTrkB, 39–46, 48, *49*, 51
Opto-YAP, 117, *118*
oximetry, 162

P

parthanatos, 88
phosphate
Dulbecco's phosphate buffered saline (DPBS), 73
phosphatase calcineurin, 4
phosphatidylinositol 3-kinase/protein kinase B (PI3K/Akt) pathway, 41
phospholipase Cγ1 (PLCγ1)/Ca^{2+} signaling, 41, *45*
phosphoinositides, *15*
phosphoinositide-3-kinase (PI3K), 42
PM-resident phosphoinositides (PIPs), 5, *6*, 13
phosphorylation, 39
Akt, 50, *52*
ERK, 114, 117
ERK1/2, 50, *52*, 76, *77*
phospho-ribosomal S6 kinase, 61
phosphor-p70S6K, 62, *63*
photoactivatable Flp recombinase, 124, 129
photolithography, 154
photoluminescent agents, 167
photolyase-homology region (PHR), 4, 26, 38, 56, 113
photo-sensitive degron (psd), 100, 103–107
overview, **101**
photosensitive transcription factor (psTF), 101, *102*, 103, 104
pickpocket protein (ppk), 60
ppk-CD4tdGFP, *63*
ppk-Gal4, 60, *61*, 64
plasmids, 125–126
constructing, optical activation of TrkB signaling, 39–41
CreC60, 141
lentiviral transfer, *75*
lentivirus, 74
Lyn, 79
oligonucleotide sequences to generate CDC20 as target gene, **105**
Opto-TrkA, *80*
PM-resident phosphoinositides (PIPs), 5, *6*
negatively charged, 13
PPIs, *see* protein-protein interactions
programmed cell death (PCD)
remote control of, 85–94
see also, apoptosis; necroptosis; pyroptosis

programmed cell death manipulation, *87*
protein-aggregated-related diseases, 112–114
protein-protein interactions (PPIs), 38
  optogenetic clustering to map out domains required for, 7–10
  optogenetic dissection of protein self-oligomerization and, 10–12
pyroptosis, **87**, 88
  localized, 94
  optogenetic control of, 91–93

R

Ras/MAPK (ERK) cascade, 116
RIPK1-RIPK3-MLKL cascade pathway, 89–91
RTK receptor, 114, 116

S

Saccharomyces cerevisiae
  control of protein levels in, 99–107
serial peripheral interface (SPI) protocols, 159
Skinner box experiment, 167
SOCE, *see* store-operated calcium entry (SOCE) pathway
sono-optigenics, 167–171
spinal cord, 152, *153*
spinal cord injury (SCI), 55
spinal cord stimulation, 154, *155*, 161
SsrA peptide, 116
SsrB peptide, 116
stem cells
  hair-follicle-derived, 79
  neural differentiation, 71–82
stromal interaction molecule (STIM)
  CC1-SOAR autoinhibition, 9, 11, *12*
  CRAC channel with, 25–26
  functional differences in STIM variants 14–15, *16*
  monSTIM1, 124, 129
  OptoSTIM1, *8*
  STIM1-induced ORAI1 activation, 30
  STIM1-inspired optogenetic tools, 2
  STIM1-ORAI-mediated SOCE, 4–7, 10
  STIM1-Organelle interactions with optogenetic approaches, 13–14
store-operated calcium entry (SOCE) pathway, 2
  optogenetic mimicry of molecular steps involved in, 4, 5, *6*
  STIM-ORAI coupling during, 10
synergistic optogenetic multistep control (SOMCo), 104–107

T

tamoxifen, 136
tdTomato transgenic mice, 143–147
tetracycline, 136
tetramer state, 117
tetR-*As*LOV2 construct, 104–105, 107
transgenic flies, 56, 60, 65, *66*, 67
transgenic mice, *see* tdTomato transgenic mice
tropomyosin receptor kinase (TrkA)
  light stimulated, 81, 82
  neurotrophic ligand binding to, 73
  see also, OptoTrkA
tropomyosin receptor kinase (TrkA)-mediated signaling, 73
tropomyosin receptor kinase B (TrkB) signaling, optogenetic activation of, 37–51
  constructing plasmids for, 39–41
  downstream signaling activation for, 41–45
  fluorescent reporter assay, 38, 41–45
  PC12 neurite growth assays, 46–49
  Western blot, 38, 50–51

U

UAS-optoRaf, 60, 64
upconversion nanoparticles (UCNPs), 93, 163–167, *168*

V

Vaucheria frigida, LOV domain of (VfAuLOV), 38
ventral tegmental area (VTA), 167, *168*

W

Western blot
  optoRaf and optoAKT, *59*
  optoTrkA, *81*
  pERK and pACK, 57
  TrkB signaling, 38, 50–51
wireless and non-invasive optogenetics
  mechanoluminescence, 151, 167–171
  tools and technologies 151–172
  upconversion nanomaterials, 163–167
  wireless active optogenetic devices, 158–161

wireless multimodal optogenetics, 161–163
wireless passive optogenetic devices, 152–158
wireless optogenetic control of cell death *in vivo* 93–94

Y

YAP, *see* opto-YAP
yeast
budding yeast, light-sensitive protein synthesis in, 101–103
SOMCo of, 104, *106*
wild type, 105
Y-maze experiment, 167

Z

zebrafish, *see Danio rerio* (zebrafish)

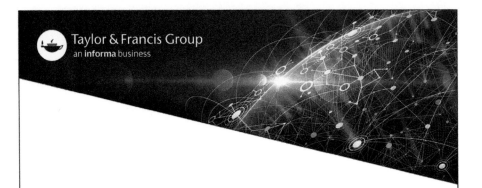

9781032249223